CAMBRIDGE LIBRARY

Books of enduring scholarl

Life Sciences

Until the nineteenth century, the various subjects now known as the life sciences were regarded either as arcane studies which had little impact on ordinary daily life, or as a genteel hobby for the leisured classes. The increasing academic rigour and systematisation brought to the study of botany, zoology and other disciplines, and their adoption in university curricula, are reflected in the books reissued in this series.

Memorials of Sir C.J.F. Bunbury

Sir Charles James Fox Bunbury (1809–86), the distinguished botanist and geologist, corresponded regularly with Lyell, Horner, Darwin and Hooker among others, and helped them in identifying botanical fossils. He was active in the scientific societies of his time, becoming a Fellow of the Royal Society in 1851. This nine-volume edition of his letters and diaries was published privately by his wife Frances Horner and her sister Katherine Lyell between 1890 and 1893. His copious journal and letters give an unparalleled view of the scientific and cultural society of Victorian England, and of the impact of Darwin's theories on his contemporaries. The final volume covers the years 1884–6. Bunbury's health was declining, but he kept up his journal- and letter-writing (although many of his contemporaries had predeceased him), and had many visitors. The volume ends with tributes written by his many friends.

Cambridge University Press has long been a pioneer in the reissuing of out-of-print titles from its own backlist, producing digital reprints of books that are still sought after by scholars and students but could not be reprinted economically using traditional technology. The Cambridge Library Collection extends this activity to a wider range of books which are still of importance to researchers and professionals, either for the source material they contain, or as landmarks in the history of their academic discipline.

Drawing from the world-renowned collections in the Cambridge University Library, and guided by the advice of experts in each subject area, Cambridge University Press is using state-of-the-art scanning machines in its own Printing House to capture the content of each book selected for inclusion. The files are processed to give a consistently clear, crisp image, and the books finished to the high quality standard for which the Press is recognised around the world. The latest print-on-demand technology ensures that the books will remain available indefinitely, and that orders for single or multiple copies can quickly be supplied.

The Cambridge Library Collection will bring back to life books of enduring scholarly value (including out-of-copyright works originally issued by other publishers) across a wide range of disciplines in the humanities and social sciences and in science and technology.

Memorials of
Sir C.J.F. Bunbury

VOLUME 9: LATER LIFE PART 5

EDITED BY
FRANCES HORNER BUNBURY
AND KATHARINE HORNER LYELL

CAMBRIDGE
UNIVERSITY PRESS

CAMBRIDGE UNIVERSITY PRESS

Cambridge, New York, Melbourne, Madrid, Cape Town,
Singapore, São Paolo, Delhi, Tokyo, Mexico City

Published in the United States of America by Cambridge University Press, New York

www.cambridge.org
Information on this title: www.cambridge.org/9781108041201

© in this compilation Cambridge University Press 2011

This edition first published 1893
This digitally printed version 2011

ISBN 978-1-108-04120-1 Paperback

MEMORIALS

OF

Sir C. J. F. Bunbury, Bart.

EDITED BY HIS WIFE.

THE SCIENTIFIC PARTS OF THE WORK REVISED BY
HER SISTER, MRS. LYELL.

LATER LIFE.

VOL. V.

MILDENHALL:

PRINTED BY S. R. SIMPSON, MILL STREET.

MDCCCXCIII.

1884.

JOURNAL.

January 1st.

I must begin the New Year with a devout ex- 1844. pression of our thankfulness, for a release from painful anxiety. Edward was on his way back from Algeria, and according to his last letter he expected to arrive at Marseilles by steamer on December 23rd, and to eat his Christmas dinner with the Colebrookes at Cannes. But we have heard nothing from him since he left Africa, and were becoming very anxious — Fanny especially. At length she telegraphed to Cannes, to Cecil, and received an answer from him that he had seen his uncle Edward there yesterday—quite well, thank God.

Truly, the telegraph is a great blessing.

January 3rd.

Miss Wood—Harry's betrothed—and her mother arrived in the late afternoon of the 1st: the formidable ceremonies of introduction have been gone through, and we are now able to be at our ease with them. Our impression—of the mother especially—is very favourable. Mrs. Wood is a thorough lady, gentle, kind, of pleasing manners, and I think also of cultivated mind (yes, certainly).

1844. Of the young lady I cannot yet judge so decidedly, but she is very pleasing. At first I did not think her pretty (which, indeed, is more Harry's concern than mine), but the more I have been in her company, the more favourable has been my opinion of her personal appearance. In more intimate and confidential conversation, on important subjects, Fanny has found both of them very satisfactory.

To return to my retrospect of last year:—

I did not find much, in our two months in London, that was very interesting at the time, or that has left a deeply agreeable impression on my memory. But I must except our visit to Lady De Ros, a most interesting and charming old lady;—so old, that she had danced at the ever memorable *Brussels Ball*, on the 15th of June, 1815, and talked with the Duke of Brunswick a few hours before he was killed. She was intimately acquainted with the Duke of Wellington. But at this great age, Lady De Ros is in perfect possession of all her faculties: she talked admirably well, telling us in the clearest and most lively way many interesting and amusing stories of the Great Duke, and other famous men with whom she had been familiarly acquainted.

Mrs. Swinton, Lady De Ros's daughter, is also an agreeable person. She spent five days here with us in October. Lady De Ros's age is not given in Debrett, but she is 3rd daughter of the 4th Duke of Richmond, therefore sister of that Duke who was in the Peninsular battles and was dangerously wounded at Orthez, of whom I have so often heard Sir George Napier talk.

1884.

We paid, from London, two very pleasant country visits to dear Kate Hoare at Minley Manor, and to Lady Grey at Fairmile, near Cobham: but they were so nearly repetitions of our experiences of last year, that I need not say much more about them.

I have noted in my journal of last year the completion of the printing of my "Botanical Fragments," and the gratifying acknowledgments I have received from my friends to whom I sent it, and especially from such men as Bentham and Hooker. I can now add another very eminent botanical authority—Dr. Asa Gray, who in acknowledging his reception of the book, not only speaks of it in flattering terms, but expresses a wish to notice it, himself, in an American journal, to which, of course, I have consented, although the book is not, strictly speaking, published. It has indeed been received more favourably, more kindly than I expected, and I fear somewhat better than it deserves. It is a comfort to me to have lived to bring it out and to place it in the hands of my friends and of those who are interested in the subject: yet I sometimes almost wish it were to be done over again, as I think I could improve it.

Received from Mr. Grant Duff, from Madras, a courteous letter of acknowledgment for my book.

The labourers' supper party.

Friday, 4th.

Weather very mild, but foggy and damp. Walked with Fanny and Mrs. Wood to the girls' school and through the arboretum.

1884.   Satisfactory confabs. The women's feast. Mr.
and Mrs. Lawrence Jones dined with us.

January 5th.

Mrs. Wood and Laura and Harry went away.

January 6th.

Read Kingsley's sermons on Self Righteousness.

January 9th.

Poor dear Car. thrown while riding in the park,
and much hurt, but—thank God—not so seriously as
we at first feared.

Mr. Wood, Mr. Henry Nicholl and Harry arrived.

January 12th.

We have lately had a sad alarm. Our dear
Car (Caroline MacMurdo) a most charming and
excellent girl, and very dear to us, was thrown
while riding in the park on the 9th, and much hurt
—though happily the injury has turned out to be
less severe than was at first apprehended. The
first report brought to us was indeed very alarming,
and gave Fanny especially, a terrible shock; Car
had fallen on her head, and it was reported that
blood was coming from her ears. The truth was
that her *external* ears were rather severely cut, as
well as her chin ; but thank God, she was not
seriously hurt in any way ; and though not able to
leave her room till to-day, she is rapidly recovering.

We may indeed be very thankful to the Almighty 1884. that no worse has been the result of an accident which terrified us so much at first. Besides the risk of more serious harm, we may well be thankful that her lovely face (truly lovely) has escaped any disfigurement.

In the evening of the 9th, Mr. Thomas Wood (Laura's eldest brother), Mr. Henry Nicholl (one of the firm of our lawyers) and Harry arrived. On the 10th we two held a serious consultation with these three and Thomas Scott, to settle the business of the settlements &c., required for the marriage of the young people. The business was tedious and tiresome, but was conducted in a very amicable spirit throughout, and brought, I trust, to a satisfactory conclusion.

January 13th.

Confined to the house by a cold,—read prayers with Fanny, and read Kingsley's sermon on the Love of God. Agnes and Constance Wilson stay-with us—very pleasant.

January 14th.

I confined to the house by a cold. Dear Agnes Wilson and her sister went away.

The Barnardistons (4) the Charles Hoares and Truey Hervey, Laura Wood and her sister, the Iveses, Emmy Bunbury, and Arthur Hervey arrived —and Reginald Talbot.

1884.   January 15th.

A very fine day and mild morning.   Walked an
hour in the garden.   Saw Dr. Macnab.

All the party except Barnardiston, Car. and I
went to the Bury Ball.

January 16th.

The young people very merry in the evening
acting proverbs very amusingly.

January 17th.

Our dance—seemed to give much pleasure to the
young people.

January 24th.

We were exceedingly shocked and pained the
day before yesterday, by the news (which we first
learned from *The Times*) of poor Lord Hertford's
fall and his dreadful hurts.   We have since had
letters from both Sarah and her mother giving most
touching and pathetic details of the deplorable
event.   Sarah and Albert were actually riding with
him when the disaster happened ; and Minnie, who
was on her way from London to a visit at Ragley,
arrived just as he was brought in.   It is a most
lamentable misfortune.   I did not, indeed know
Lord Hertford intimately ; but it seems certain that
he was one of the best of men, deeply loved by all
his family and dependents, and by all who knew
him well ; and it needed but little acquaintance to
be struck with the peculiar charm of his manners,

which were not only graceful and noble, but were 1884.
characterized by unmistakable goodness of heart.
I suppose no one could be more happy in his family
—surrounded by his numerous children and grand-
children, all so much attached to him. And now
the failure of a horse has broken up all this happiness and put a stop for ever, in this world, to the
benignant influence of this good man. Truly,
indeed—" in the midst of life we are in death."

Albert had to pull the horse from off his father,
who lay crushed under it, and poor dear Sarah had
the terribly painful task of breaking the shock of
the news to poor Lady Hertford.

January 25th.

Poor Lord Hertford is still alive, but there seems
not to be the slightest hope of his recovery.
Happily, it is said, that he suffers no actual pain :
but he is paralyzed up to the chest. It seems to be
still unknown what occasioned the fall of the horse ;
most likely some sudden illness.

January 26th.

Lord Hertford was released by death yesterday
evening. As there was not the slightest chance of
his recovery, one could not wish, either for his own
sake or that of the friends about him, that his life
should be prolonged in that helpless and hopeless
state.

LETTERS.

My dear Katharine,

1884.

Many thanks for your very kind and pleasant letter, which, though it did come the day before that for which it was destined, was very welcome ; indeed, on reading it over again, I do not perceive that it is any the worse for keeping! I thank you heartily for your kind wishes, and entirely agree with you that we can only be thankful for our blessings, and pray that our dear ones may be preserved to us and to one another. I am sure I feel that I can never be grateful enough to God for all the blessings which have been bestowed on me. To enjoy such good health as I do, at the age of 75 (though I am not *strong*), and to have such a wife, and such friends—these are blessings indeed. I fully trust, too, that we may feel our happiness prolonged and continued in that of Harry and his bride, for there seems every reason to hope that she is one who will contribute to the happiness of all with whom she may be connected.

Many thanks for your kind intentions in sending me "Hanley Castle," but I had it already from Mr. Symonds, and have read most of it:—it is well written, but I think it is too much crowded with incidents, and the interest not sufficiently concentrated.

I am very sorry to hear that Mr. Symonds is

suffering so much, and also that the Hookers are in so much anxiety about Mrs. Dyer, and that Mr. Bentham is worse. I daresay the variableness of the weather is trying to persons in delicate health. But it is really an extraordinary season ; our garden is full of snowdrops, crocuses and star-anemones, and hepaticas and blue squills are beginning to appear, and honeysuckles are in leaf. I fear they will be cut off sooner or later by sudden frosts. I am glad you have time now and then to return to your ferns.

A curious plant has lately been sent to me by Lady Lilford :—the Mexican Selaginella which rolls itself up into a ball when dry, and revives when wetted. It was sent with the name of *Selaginella rediviva*, but I think it is the same which is figured by Sir William Hooker in the second volume of "Icones Plantarum," under the name of *Lycopodium lepidophyllum*.

I have not time to write more just now: but with love to Rosamond, believe me ever,

Your loving brother,

CHARLES J. F. BUNBURY.

————

Barton Hall,
February 7th, 1884.

My dear Leonora,

I thank you very much for your pleasant letter which I received to-day, and for your very kind wishes concerning my birthday. I may well be very thankful for enjoying such good health at the age of 75, and still more for having so many dear

1884. and valued friends. Long life is not *always* a thing
to be wished for, but when it is attended by such
blessings as these, one may indeed be grateful for it.

Your letters and Susan's give a most pleasant
idea of the life you are all leading; so active and
cheerful, with so many various occupations, all good
and useful as well as agreeable. I am very glad to
hear that Dora is continuing her botanical and
geological pursuits. I can better appreciate these
than her mathematics, that subject having been
as little to my taste as to Alfieri's. I am very glad
also to hear of the young ladies' success in the
ambulance examination.

We have hitherto, in this country also, a remarkably mild winter : the wood-pigeons are cooing
as if it were May, and it is delightful to see the
quantity and variety of flowers in the garden :—
Snowdrops, two species ; Anemone pavonia ; He-
patica ; Scilla bifolia, and white Arabis,—besides the
Jasminum nudiflorum. We are just at present
no more than a quartett, Car. and Arthur Mac
Murdo being the only additions to our conjugality ;
and I am thankful to say we are all quite well.
Dear Car. has quite recovered from the effect of her
fall, and is as lovely as ever ; she is really a
charming and loveable girl, and we both are very
fond of her. She and her brother are very happy
together. Lord Hertford's death was indeed a
terrible shock to Sarah and Albert, and all who
were on the spot, and is felt as a very lamentable
event by all who knew anything about him ; I
myself knew him but slightly, but there was some-

thing about him remarkably winning, and from 1884.
all I have heard, I believe he was one of the
best of men and exceedingly loved by all who
were in any way connected with or dependent on
him.

Since I began this letter, I have received dear
Susan's and Joanna's of the 4th, and instead of
merely sending my love, I must write a few lines
to them.

Believe me, with much love to Annie and Dora,
   Ever your loving brother,
      CHARLES J. F. BUNBURY.

          Barton.
         February 8th, 1884.

My dear Susan and Joanna,

   I hope you will pardon me for writing you
only a joint letter, as I really do not know how I
should find matter for two separate letters, while
our lives are so uneventful, and Fanny keeps up
such a copious correspondence with you.

I thank you both very heartily for the kindness of
your congratulations on my 75th birthday, and your
good wishes for the time that may remain to me.
Many returns of a birthday, at my time of life, are
not to be expected, and perhaps not to be wished
for ; but I feel very grateful for what has been
granted to me, and try to be hopeful as to the
future—I mean as to the remainder of this life. I
do not mean with reference to politics, for, as to
them, the prospect looks very dark indeed.

We may, I trust, look cheerfully and hopefully to
the near future of our own family, for we are very

1884. much pleased with the wife of our nephew Willie, and not less so, as far as we can judge, with Harry's bride. What a goodly assemblage of children and grand-children Katharine has gathered round her!

You seem to be immensely industrious and pains-taking with your new book on Florence, and I have no doubt it will be very interesting and very valuable. I should like very much to study it. The only book (excluding botany and geology) that I have in hand at present, is Green's " Making of England." By this odd title he means the conquest of Britain by the Anglo-Saxons as they have generally been called (the Engles and Saxons, as he, following Mr. Freeman, calls them) and their gradual amalgamation into one people. It is rather stiff reading. I have no doubt you find Froude very interesting, and certainly he has a marvellous power of narrative, but I think one feels also that he is very much of a partisan or advocate. You will know from Fanny's letters that dear Car. Mac-Murdo has quite recovered from the effects of her fall, and is not at all disfigured. It is a great mercy, for the accident seemed at first to be (and might have been) a very bad one. We are alone at present with her and Arthur, but we expect a large party next week, and among them Laura Wood and *two* of her uncles. We had a pleasant party last week—Lady Grey in particular; but we were disappointed of the Freres, as poor Sir Bartle became and still is dangerously ill—though there seems to be now some slight improvement in his condition.

We cannot boast here, as you do, of our 1884.
cloudless, blue skies, but the winter has hitherto
been very remarkably mild, and on the whole
but little rainy. The garden is quite beautiful with
out-door flowers, as I wrote to Leonora.

Now, with my best love to you and yours,
believe me ever,

Your loving brother,

CHARLES J. F. BUNBURY.

P.S.—With respect to the question which you
have asked us in your last letter, Fanny and I
both think that there can be no objection to your
dedicating your new edition to Princess Mary. She
seems to be really clever, and to have a genuine
love and appreciation of art, so that she is worthy
of the compliment—for a real compliment it will be
in my opinion.

JOURNAL.

February 18th.

Since the last entry in this journal, we have had
very pleasant company: the first *batch*—Lady
Grey, Minnie Powys, the Leckys and Edward
(besides a few others), from January 29th to 5th
of this month. Lady Grey always interesting and
admirable, and a most true and constant friend.
Minnie Powys had not been here for several
years, owing to ill-health, but now seems quite
restored, quite her former charming self, and

1884. still very pretty.   Mr. Lecky is full of knowledge :
his conversation full of matter.   The second party
of guests, who began to come on the 12th, and who
are partly still here :—Lady Mary Egerton and her
daughter May, Willie and Mimi Bruce, Sir David
Wood here from 12th to 15th, Sir Alexander and
Lady Wood from 16th to 19th, Mr. and Mrs. John
Marsham, Mr. Lacaita, Katharine and Rosamond,
Mr. Montgomerie, Laura Wood, Harry, Lord
John Hervey, also Lord and Lady Rayleigh from
16th to 18th.

The 4th of February was my 75th birthday ; I
thank Almighty God most humbly and heartily for
permitting me to arrive at such an age, in such good
health, and in the enjoyment of so many blessings.
Little could it have been expected in my childhood
that I should live to see my 75th birthday.

——————

February 20th.

All our guests now gone except Mimi Bruce,
Car. and Arthur.

The John Marshams are a remarkably interesting
and pleasing couple—*she* more particularly so.

On further intimacy with Laura Wood, our
favourable impression of her is strongly confirmed ;
we have now not merely liking but love for her.   Sir
David Wood (one of Laura's uncles), seems a warm-
hearted, affectionate man, very much depressed by
the recent death of his wife : very much attached to
his niece.

Sir Alexander Wood (another uncle of Laura), 1884. good-natured, very fond of Laura, and very cordial to us on her account.

Mr. Lacaita, a very intelligent and remarkably well informed young man, of very pleasing manners: a capital botanist. I have learned a good deal from him about the botany of Greece and Italy.

-- ——-

February 21st.

I seldom notice in this journal any of the political events of the day, which are so amply discussed *(usque ad nauseam)* in the newspapers: and the melancholy and shocking events in the Soudan have by no means encouraged me to deviate from my rule. But I cannot avoid saying something of the news which came yesterday, of the achievement of that splendid character General Gordon. If there is no reverse, if he goes on as successfully as he has begun at Khartoum, and is able (as seems now to be expected) to bring the war to an end and to pacify the Soudan without further bloodshed, it will be one of the most admirable exploits recorded in history. He seems to be indeed a true Christian Hero.

———

February 22nd.

We had the news in a letter from Cissy, that my nephew George is engaged to be married — in Canada—to a Miss Cottle, the daughter of a widow. He has been quick in following the example of his brother Harry. If he has found a bride as likeable

1884. and loveable as Laura Wood he is fortunate indeed. And at any rate he has probably done wisely ; as no doubt the dreariness of a Canadian winter will be much more endurable with a wife. Dear Cissy is very much pleased.

February 27th.

Dear Car. MacMurdo left us yesterday (with her brother Arthur) to join the William Bruces at Rose Bank, and to remain there till after Harry's marriage. After which we hope she will return to us. Mimi Bruce stayed with us from the 16th to the 25th ; her husband spent two Sundays, the 17th and 24th, and one Monday, the 18th, with us, being able to return to his business in London for the remainder of the time. The two sisters, Car. and Mimi (Caroline and Emily) are both of them delightful, very clever and very good. Car.'s staying with us this winter, ever since October, has contributed very much to our cheerfulness. Both are lovely ; Car. has the more regular beauty, but her sister has something wonderfully interesting and attractive in her face. Mimi's baby, William Fox, born last August, is a very fine little fellow. Of William Bruce I have as high an opinion as I had last year : I do not know any other young man who comes near to him in the combination of moral and intellectual merit.

February 28th.

I read in the newspapers, two or three days ago, the death of Milner Gibson. He was only two

years older than I : I thought he had been more. 1884.
For some years, between '35 and '38, I used to
meet him frequently when I was a zealous
politician, and he was one of the conspicuous
members of the "philosophical Radical" party, with
Grote and Sir William Molesworth. But for many
years past I have seen him but seldom : indeed he
has lived much abroad. He was a cheerful,
pleasant man. He dined here with us on the 10th
of October last : he seemed then in good health,
though very deaf.

February 29th.

Within the last month (I have forgotten to note
the day), I noticed in the papers the death of Dr.
Balfour, the well-known Professor of Botany at
Edinburgh. Latterly (if I understand rightly), he
had resigned the Professorship, on account of his
age and health. He was, I believe, an excellent
lecturer, and altogether a very efficient and useful
Professor, but by no means in the first rank as a
systematic or a general botanist. He was very civil
to me when we were living at Edinburgh in the
winter of 1849-50, and gave me many specimens of
rare Scottish plants.

The weather for these last two months has been
very remarkably mild for winter. In the whole of
January there were only 7 days (or nights) on which
the thermometer was as low as freezing point.

Snowdrops and "Aconites" (Eranthis) were in
blossom before the middle of the month : and by
the end of it there were, besides these, Crocuses

1884. of two species, Mezereon, Forget-me-not, Anemones and Sweet Violets in plenty. In February, the number of frosty days (those on which the thermometer was down to or below 32 deg.) was 13 ; and the lowest point to which it descended was 25.

Our garden has been very gay with spring flowers all through the month ; rose bushes, honeysuckles and the ornamental sorts of clematis are part expanding their young leaves, and there has been no frost sufficient to hurt them.

March 3rd.

Money business with Fanny and Scott.

Wrote to Dr. Asa Gray. Arthur arrived from London.

The news of the terrible battle fought near Suakin, on the coast of the Red Sea, between our forces under General Graham, and the Arabs under Osman Digna. Our loss, though not positively great, is considerable in proportion to our numbers ; it is evident from all the particulars given, that the Arabs fought with desperate valour, and that in spite of their inferiority in discipline and in weapons —they are by no means a contemptible enemy. I am afraid the war is not at an end and if it should go on, I fear the force we have there is not sufficient.

March 4th.

Mr. Henry Nicholl and Harry came from London bringing the draughts of deeds concerning Harry's

marriage — discussion of same — all comfortably 1884.
settled.

March 6th.

Harry Sanford and Gerald Campbell came.

Mrs. Wilson and three of her daughters came to
tea—very pleasant. Constance and Ida remained.

March 8th.

Received from Messrs. Nicholl, the two important
deeds (parchment) connected with Harry's marriage
—signed them, Scott being witness, and returned
them.

March 10th.

Dear Arthur set out for London, his leave being
nearly at an end.

Wrote to Laura Wood.

March 11th.

Archdeacon and Mrs. Chapman and two daughters arrived.

My nephew Harry was married to Laura Wood,
this day at St. Peter's, Eaton Square. We had a
telegram announcing the event to us, a little before
1 p.m. So that business is settled. God grant
that it may indeed be a happy event (as there
seems every reason to hope that it will be) happy in
itself and in its consequences, now and hereafter, in
this world and in the next. I am a great advocate
for marriage, and have long wished that Harry

1884. should find a suitable wife. I think he is of a character to make a good husband, and I feel convinced that a happy marriage will contribute more than anything else to his virtue and happiness. I believe that Laura will be a very good wife ; all that I have had the opportunity of jndging of in her I like very much ; and so does Fanny, who is a much better judge.

After I had written the last paragraph, we received a telegram from the new married couple, safely arrived at the Lord Warden Hotel at Dover.

Another frightful battle in the Soudan ; and this time we very narrowly escaped a defeat, which would have implied the total destruction of the force under General Graham. One of the squares in which our troops were formed was actually broken by the enemy, and men of the different nations mixed up together, so that a great many of our men were killed in close fight by swords or spears. Our loss of brave and noble officers and soldiers, was lamentable, but it is frightful to think of the disaster it was near being. It would have been the greatest catastrophe that has befallen our armies since the first Afghan war. And though of course one feels the most for our own people, I cannot help being sorry for the awful slaughter that has been made of the brave Arabs, who need not (so far as I can see) have been our enemies. Brave and gallant enemies they certainly are ; but why should we send our troops to slaughter or be slaughtered by them ? It appears to me a cruel, useless, and lamentable war.

Letters about the wedding.

Had a pleasant walk with Agnes Wilson (who arrived after breakfast), and one of the Miss Chapmans.

Fanny went into Bury with the Chapmans, to a meeting of the Prevention of Cruelty to Animals Society.

Lady Hoste and Dorothy and Mr. Green dined with us.

————

March 13th.

The Chapmans went with Fanny to a meeting at Bury, about a new school for girls—they went away from hence—dear Aggie Wilson remained with us.

————

March 14th.

Went out with Fanny and Aggie in the open carriage to Ampton. We saw the John Paleys.

Dear Agnes Wilson went away.

————

March 15th.

A most beautiful day, quite summer.

Fanny took me out in the pony carriage to Fornham St. Martin, and back by another way.

Dear Minnie and dear Car. arrived from London.

————

March 16th.

We went (Minnie and Car. with us), to morning Church and received the Communion.

1884.   March 17th.

Had a pleasant drive in the open carriage with Fanny and Minnie,—was out for two hours.   Dear Mrs. Wilson and Aggie came to tea.

———

March 18th.

We went with Scott for nearly two hours, through the groves and plantations inspecting and marking trees.

Fanny has had a delightful letter from Laura and Harry, full of love and happiness.

———

March 20th.

Lady Hoste and Dorothy came to luncheon and stayed to tea.

———

March 24th.

Death of our dear old dog Harold.

———

March 25th.

Scott came in from the Vestry Meeting and reported all satisfactory.

———

March 28th.

The sad news of the Duke of Albany's sudden death.

A visit from Mr. Harry Jones.

———

March 31st.

Scott's comfortable account of Rent Audit at Mildenhall.

[During the early part of March the weather was 1884.
beautiful and summer-like, but the latter part was
cold and disagreeable. We were both suffering
from colds, but were not much confined to the
house. Sir Charles was writing his Reminiscences,
and some Notes on Wild Plants, and reading the
"Life of Mountstuart Elphinstone," Mrs. Roundell's
"History of Cowdray," and Collin's "Sophocles."—
F. J. B.]

April 2nd.

A beautiful day with a strong S. wind. Went out
with Fanny in the open carriage, and afterwards
walked with her in the garden.

April 3rd.

A splendid day.

My Barton Rent Audit—satisfactory; tenants
very civil: luncheon afterwards. Fanny, Minnie
and Mr. Harry Jones joined us in it.

April 4th.

Dear Car. laid up with the *measles*. Walked with
Minnie. Mr. Bevan and his daughter Mabel came
to tea.

April 7th.

The master of Trinity and Mrs. Thompson came
to luncheon. Dear Car. going on well. Received a
letter from dear Charlie Seymour, and wrote to him.
Sent my book to Miss North and to Lord Walsingham.

1884.   April 9th.

Had a pleasant walk with Fanny (*she* in the pony-carriage), along the green ride.

Had a pleasant note from Miss North in acknowledgment of my book.

April 10th.

Had an agreeable note from Lord Walsingham in acknowledgment of my book.

Clement arrived.

April 11th.

The 40th anniversary of our happy betrothal—thanks be to God.

Read prayers with Fanny and Minnie ; we did not go to Church because of the prevalence of measles.

April 21st.

Lord and Lady Rayleigh, Mr. and Mrs. William Hoare and Mrs. John Paley came to luncheon.— Lady Galway and Miss Waddington afterwards.

April 23rd.

Received a pleasant letter from Mr. Grant Duff, with a present of seeds of Indian plants.

Dear Car. allowed to come down to luncheon with us :—much rejoicing.

April 24th.

Mr. and Mrs. William Hoare and John Herbert arrived—a pleasant evening with them.

The *earthquake* of Tuesday, the day before 1884. yesterday, appears to have been the most serious that has been felt in England (especially in this part of England) for a long time. Its centre of action seems to have been about Colchester and the neighbouring villages; there the damage was very serious indeed—cottages and school-houses wrecked, large chimneys and even steeples thrown down, church walls cracked; the newspapers are full of the details. *Here* it was felt, but not severely; Scott was writing in his office about half-past 9, when he suddenly heard a loud and strange noise, unlike (he says) to anything he remember to have heard before, and instantly afterwards (or almost at the same time) his head was affected by a strange, indescribable, uncomfortable sensation, so that he thought he was going to have a "fit:" but it soon passed away. Neither Fanny or I felt it, but Car. MacMurdo did.

At Bury it seems to have been very generally and decidedly felt, though it did no positive damage.

April 25th.

Walked in the garden with Fanny and the William Hoares. Dear Car. quite recovered. Mr. James and Mrs. Victor Paley dined with us. The William Hoares both sang delightfully.

April 26th.

A beautifully mild day. The W. Hoares went away.

1884.

April 29th.

Had a very pleasant walk with Minnie and Car.
Lady Hoste came to tea.

_____

April 30th.

Agnes and Constance Wilson arrived.   Dorothy
Hoste came to tea.

Examined and verified species of a Cape grass in
my collection.

_____

May 1st.

Harry and Laura arrived from their wedding tour
—a hearty welcome.

Had a pleasant walk with Agnes Wilson.

_____

May 2nd.

Had a pleasant walk with Laura.   Mrs. Wilson
came to luncheon.

_____

May 5th.

Mr. and Mrs. Livingstone arrived.

_____

May 7th.

The Livingstones went away in the morning.
John Herbert in afternoon.

_____

May 8th.

A beautiful mild day.

Fanny, Laura and Harry spent the day at Mildenhall, returning to dinner.

Gave my "Botanical Fragments" to Laura—sent another copy of same to Leonard Lyell.

May 10th.

A most beautiful day, very hot. Had a very pleasant drive in pony carriage with Fanny through the home farm green drive.

Our dinner party:— Mrs. Wilson and two daughters, Lord John Hervey, Mr. Quayle Jones, the Victor Paleys, Mr. J. J. Bevan and his daughter, Mr. and Mrs. Algernon Bevan.

May 12th.

Dear Harry and Laura went away early. Drove out with Minnie and Fanny. We called on Miss Waddingtons and Patrick Blake.

May 13th.

Dear Minnie and dear Car. MacMurdo left us after luncheon ; we strolled through the Vicarage Grove before with them, and we were alone for a wonder. We arranged a great many books.

May 14th.

Weather unsettled, rather stormy and chilly. Mr. Lockwood, the poor law Inspector, came to luncheon, and we discussed with him and Scott the questions of the Mildenhall Workhouse.

LETTER.

Barton Hall, Bury St. Edmund's,
May 14th, 1884.

My dear Edward,

1884.

We have settled to go to Town on Monday next, the 19th; we should not go so soon, only Kate Hoare has invited us to dinner on the 20th, to meet her father and mother, who are going down to Wells almost immediately after, and we cannot miss the opportunity of seeing them. We have been enjoying some delicious days for nearly a week past, and I never enjoyed more the spring beauty of Barton : but now the weather seems to be changing, which rather mitigates my unwillingness to leave home.

We have had an extremely pleasant *home* party :—Harry and Laura, Minnie and Car. Mac-Murdo, but now they are dispersed, and we are (for a wonder), quite alone, though expecting soon to meet them all again. I have grown very fond of Laura, as I have been, much longer, of Car.

As you have not written anything to us about the *earthquake*, I presume you did not feel it. Neither did Fanny nor I, nor (I believe) anyone in this house except Car., and she not much ; but Scott, who was writing in his office, felt his head considerably affected for some moments, and heard a strange unusual noise. At Bury it seems to have been felt pretty generally, though it did no considerable damage. You will of course have read the newspaper accounts of the sad havoc it caused in 1884. Essex.

We are both well, I am thankful to say, as I hope we shall find you when we are in London.

Ever your affectionate brother,

CHARLES J. F. BUNBURY.

JOURNAL.

May 17th.

A business meeting in this house of Trustees of Poor's Firing Farm.—Mr. H. Jones, Mr. King, Mr. Denton, Mr. Baldwin, Scott and myself—a long and laborious discussion.

May 18th.

We went to morning Church : heard an excellent sermon from Mr. H. Jones, and received the Communion.

[During the month of April and the beginning of May, Sir Charles was writing his Reminiscences, and his paper on the wild plants of Barton, and reading the "Life of Mountstuart Elphinstone." The Duke of Argyle's article on Communism (The Prophet of S. Francisco). Lord Hastings' journal in India. Duc de Broglie's "Frederick II. and Marie Thérèse." Mr. G. Brodrick's paper on Democracy and Socialism.—F. J. B.]

May 20th.

Up to London by the half-past 8 a.m. train to

1884. St. Pancras: had a good journey and arrived safe
and well—thank God. We dined with the Charles
Hoares: met our dear Bishop and Lady Arthur
and Truey.

———————

May 21st.

A fine, bright day, but cold.

We visited dear Mimi Bruce: she not looking
well,—her dear little boy thriving exceedingly. We
visited Mrs. Mills also. Minnie dined with us.

———————

May 22nd.

Fanny took me to the Natural History (British)
Museum, where I spent an hour very pleasantly
while she was at Mrs. Hubbard's party.

A visit from Mr. and Mrs. Hutchings; we visited
Mrs. Sancroft Holmes.

———————

May 23rd.

Cousin Catty, Cissy, Emmie and Sally came to
luncheon. Lady Arthur and Kate Hoare directly
after.

———————

May 24th.

Mrs. Mills and the Clements Markhams lunched
with us: William afterwards went with us to
Wressil Lodge, Wimbledon, where we visited the
Miss Freres—they very cordial—the place looking
beautiful.

———————

Read prayers with Fanny, who was not quite well. Read Stanley's fine sermon on the Dedication of Westminster Abbey.

May 26th.

Drove with Fanny: we admired the Rhododendrons and Azaleas in the Park. Gave directions about binding of books.

Our dinner party:—Bishop of Bath and Wells, Kate and Charles Hoare, Sally, Car., Dora Walrond, Minnie, Cecil, Harry Bruce—very pleasant.

May 27th.

We went (Minnie with us) to see Bull's exhibition of Orchids—magnificent.

Willie and Mimi Bruce dined with us—both delightful.

May 28th.

Katharine and Mary Lyell and dear little Charlie came to luncheon.

Our dinner party :—the MacMurdos, Mr. Robert Marsham, Clements Markham and his daughter, the Walronds, Minnie, John Herbert, Edward— pleasant.

May 29th.

At 48, Eaton Place—to go back a little.

Harry and Laura arrived at Barton—from their wedding tour, on the first of this month. They had

1884. a warm welcome : not only from us, but the people
of the village, as well as of the household, had
assembled to greet them, with ornamental arches,
flowers, music, &c. They remained with us till the
12th, and it was a very pleasant time ; they both
were in excellent spirits, very loving and happy. I
grew very fond of Laura, and I cannot be sufficiently
thankful that Harry has been so providentially
guided in his choice of a wife. I have been very
anxious, and so has Fanny, that he should marry,
and marry one who would be suited to him, being
sure that his future conduct and happiness would
mainly depend thereon ; and now we may well
thank God that he has found a wife who seems to
be everything that we could wish.

Dear Minnie Napier and Car. MacMurdo left us
on the 13th, and we were alone (for a wonder) till
we came up to London on the 20th. Car. has been
living with us ever since October last, and I was
very sorry to part with her. I love her dearly, and
she is indeed a most loveable creature : not only
beautiful in a remarkable degree, but very clever,
lively, merry, full of drollery and gay spirits, at the
same time very affectionate, and (as far as I have
ever perceived) sweet tempered, though high
spirited. She has frequently been alone with us
two old people for a considerable time together, yet
she has never (so Fanny assures me) complained of
dulness, nor shown any symptom of weariness or
ill-humour ; she has always been cheerful and
obliging, and to me her behaviour has always been
charming.

Most part of this month, before we left home, 1884. was very fine, and the garden and park and groves of Barton, in all their full vernal beauty, were so delightful, that I felt very unwilling to leave it. Indeed I have felt this year after year, for a good many years past, the vernal delight,—the beauty of the spring flowers — even the commonest — the delicate varied tints of the foliage, and the interest of watching the successive development of the different kinds, the notes of the birds—all this has a charm for mé which I find not at all diminished by advancing age, and I have felt very unwilling to exchange it for London.

Car. has dined with us since we came to town— on the 26th—and she was in even more than her usual beauty. Dora Walrond, who was another of the party, is a very handsome girl.

———

May 30th.

The 40th anniversary of our happy—truly happy —marriage. I can never be grateful enough to the Almighty for bestowing on me the blessing of such a wife, and allowing me to enjoy that blessing so long.

We were grieved by the news of the death of Sir Bartle Frere. He had a long and distressing illness, which must have been most harassing to himself and his family, but was not considered hopeless till within a very short time; indeed Louis Mallet, calling on him less than a week ago, found him apparently in a decidedly improved state of health,

1884. cheerful and hopeful, so that he (Mallet) congratulated him on his hopes of recovery.

Sir Bartle Frere was a remarkably agreeable man, and some years ago we saw much of him, and had great pleasure in his society. I thought him very seriously mistaken in his South African policy, when he plunged us into the Zulu war; but I do not doubt that he honestly and conscientiously believed the course he had pursued to be just and necessary. His Indian career was so honourable and brilliant, that I cannot help wishing that he had never accepted the Governorship of the Cape.

————

May 31st.

While we were quietly at dinner yesterday evening, at Katharine's, there were some more of the desperate and atrociously wicked attempts to cause extensive destruction by means of dynamite; but again, thanks be to God, the mischief done has been small in proportion to the intention. It is no thanks to the villainous plotters that the whole result of their wickedness has been to break a quantity of glass, and to wound a few poor housemaids and kitchen-maids. But it is disagreeable to live in a city where one is any day liable to be endangered by the wickedness of such infernal villains.

————

June 2nd.

Mr. Swinton, Susan MacMurdo and dear Car. came to luncheon. We drove round Regents Park

and Hyde Park—called on the Loraines, and saw 1884.
their pretty children.

Sir Bartle Frere is to be buried in St. Paul's, the
Abbey being too full for additional graves. No
wonder, considering the scale of many of the tombs
of former ages.

——————

June 3rd.

Arthur arrived from Dublin about 7.30 a.m.
Harry and Laura about 6.30 p.m. Katharine and
Agnes Kinloch came. Susan and Car. MacMurdo
dined with us.

——————

June 4th.

Sarah Craig and two of the Napier Girls came to
luncheon with us.

Minnie and John Herbert dined with us.

——————

June 6th.

The Barnardistons and Emmy came to luncheon.
Visit from Lord Tollemache. Edward dined with
us.

——————

June 7th.

Mrs. Wilson and daughter came to luncheon. A
visit in my study from Sir Alexander Wood. Minnie
dined with us.

——————

June 10th.

Mr. and Mrs. Sinclair, Montagu MacMurdo, and
dear Car. came to luncheon. I had a drive in Hyde
Park, with Fanny and Car. Lady Head came to
tea.

June 11th.

Patience Morewood and her brother Constantine came to luncheon.

We visited the National Gallery, and spent some time there with much satisfaction, renewing acquaintance with many old favourites. Noticed a few novelties: in particular, a "Christ at the Pillar," ascribed to Velasquez — a horrible and revolting, almost disgusting picture, I should call it; inspired by the spirit of the Inquisition. A very different work is a portrait of a boy, by Isaac Ostade —very natural and charming.

———

June 12th.

A beautiful and very warm day. We went, Laura with us, to Rose Bank. It was in great beauty and very enjoyable in this fine weather. The river very full — high tide — and lighted up beautifully by the sunshine. A pleasant family gathering: dear Mimi with her baby; Car in great beauty, "*Weena*" Hampson (Louisa MacMurdo), looking very far from well however, some of the Bruce girls; several other friends, a very pretty Mrs. Napier, a daughter-in-law of Lord Napier, with her beautiful little boy. Susan MacMurdo, wonderfully handsome still, at the age (she says) of 58.

———

June 13th.

Arthur returned from Ascot. Montagu and Susan and Car MacMurdo, Emmy and Minnie dined with us. Fanny went to a ball.

June 14th.   1844.

Our dinner party. Lord Cornwath, Lady Edith Adean (his daughter) Leopold and Lady Mary Powys, and their eldest daughter, the Gambier Parrys, the Palmer Morewoods, Katharine, Admiral Spencer, Mr. Hubert Hervey, Emmy Bunbury—a pleasant party.

June 15th.

Fanny went with Arthur to the early Communion, afterwards she and I went to morning service in Eaton Chapel.

June 18th.

Leonora and her two girls dined with us.

June 19th.

Our dinner party: — Dowager Lady Rayleigh, Harry Bruce and his wife, Lothian Nicholson and his wife, Arthur Lyell and his wife, William Napier, Annie Pertz, Car MacMurdo, Mr. Sanford and his daughter Rose, Edward, C. Newton, besides our four selves.

June 21st.

A beautiful day.

Went out with Fanny; first time I have been out since Sunday the 15th; for on that day I caught a bad cold in coming out of Church, and have been laid up ever since. We drove to the Botanic Gardens in the Regents Park, and strolled in them

1884. for some time, enjoying very much the beauty of
those gardens, which are in the highest order and
condition—the brilliancy of the grass, the profusion
and variety of the flowers, and the warbling of the
numerous birds. Indeed I have long admired those
gardens, which are quite remarkable for their arrangement, for the abundance of beautiful and
interesting objects brought together within the small
space, and for the delightfully fresh and rural
appearance of everything. I was struck with the
beautiful appearance of a bed of flowers, which
turned out to be nothing else than Epilobium
angustifolium.

June 22nd.

A visit from our dear Mrs. Storrs—a great
pleasure, and one which had been long delayed, for
her children have been ill with the hooping-cough
(which I have never had), and so, though we were
living within so short a distance, all communication
between the two families was cut off. At last the
embargo has been removed. She was delightful.

The same afternoon, we had a visit from Judge
Denman, who talked very agreeably. He is a man
of fine and dignified appearance and agreeable
manners. He offers to show me the new law
courts.

June 23rd.

Our dinner party :— Mrs. Swinton, Mr. and Mrs.
R. Strutt, M. and Mme. de Bunsen, Sir Joseph and

Lady Hooker, Mr. and Mrs. Andrews (American), 1884. the MacMurdos, Leonora, Edward, Harry and Laura.

————

To the Natural History Museum, but could stay only a short time. It is a delightful place. In the galleries on each side of the great hall, a number of British birds, most beautifully mounted and arranged :—a male and female of each (where the sexes differ in plumage), with the nest and eggs or unfledged young, surrounded by rushes and bulrushes, or ferns, or leafy branches of trees—in short, each kind placed in the midst of its natural accompaniments, so that at a very little distance they may seem alive. And each kind is in a separate glass case on a stand by itself, so that it may be seen from all sides.

————

We heard of the death of Lord Arran,—having heard only a very few days before of his serious illness. It cannot however be called premature, as he was 83 years old. Last year he had a very severe and dangerous illness, but he apparently recovered from it, and (I am told) appeared to be in better health than for some years before ; so that his departure at last was rather sudden. I have scarcely seen him for several years past ; but I am very sorry for his wife and children. We were cousins by marriage, as he married (in 1838, I

1884. think), my cousin, Bessie Napier.—My acquaintance with him began in 1834, when he was British Minister at Buenos Ayres.

————

June 27th.

At Barton.

We came down hither with Harry and Laura, yesterday, because of a sudden alarm of scarlet fever in our house in London; and we propose to go back after nine days, when (according to our doctor) the danger of infection will be past.

The weather is now splendid, and this place is in great beauty, so that I must confess I cannot be sorry to be here in the enjoyment of the trees and flowers and the sunshine and shade, instead of amidst the noise and glare and smoke of London. The hay-making is going on briskly, and is pleasant to look at, though the crop will probably not be large. The foliage of the trees this year is very rich and beautiful, and the rose blossoms are very fine.

A curious and beautiful plant of the lily family—the Gloriosa superba of Linnæus—is now blossoming well in our hot-house, the first time I have ever seen it alive, though I have known it by description for a long time. It was given us by Mrs. Sancroft Holmes, after her return from Ceylon. two years ago.

The grass, as might be expected after such a long drought, is looking rather white and burnt up (except under the trees), but the foliage is very fine, and the flowers also, especially the roses.

Phillips has finished his hay-making, and he thinks that the *quality* will in some degree make up for the deficient *quantity*. He and I, I take it, are about the only people in the parish who have any grass land worth speaking of

---

June 28th.

Went to a parish meeting at Mr. Harry Jones' :— met there Mr. King and Mr. Denton—afterwards Mr. Harry Jones drank tea with us.

July 6th.

Sir Francis Doyle came to luncheon—also Leonora and Annie, Minnie and Ruth Wood.

48, Eaton Place.

We returned to London yesterday ; there was a grand thunderstorm between 3 and 5 o'clock, and the air has been cooler since : but not much rain fell at Barton, where it was much wanted.

The hay-making was nearly finished, and Scott tells me that the crop is not only of the finest quality, but much better in quantity than was expected. There is enough for two good stacks. The wheat crop promises extremely well ; but the barley (Scott tells me), on some of the heavy lands, is likely to be a great failure.

July 7th.

To the Natural History Museum, and enjoyed the birds even more than the first time. Met there

1884. Lord Walsingham, who has mounted and arranged a great many of those admirable groups of birds with their nests and eggs and young.

Gloriosa superba.—I see an excellent coloured drawing of this plant in the Botanical Gallery of the Natural History Museum at South Kensington.

The Botanical collection at South Kensington is very well arranged, and probably as well suited for purposes of instruction as could be; but dried plants at any rate are by no means so well fitted for exhibition as birds or insects. A great many vegetable specimens, however,—such as the gigantic stems and leaves of Palms, Tree Ferns, Pandani and Aroids, exhibit very well, and of these there is a magnificent display in the gallery of the Museum.

Called on poor Bessy Arran (formerly Bessy Napier)—she looks *ghastly*. A sad contrast to her *eldest* sister Emily, whom I have seen this day, and who (though 4 years older, and with snow-white hair), is a real picture of healthy and peaceful old age,—one might really say of beautiful old age. Bessy was very gentle and kind, and talked with feeling of the old days of 1825, when she was at Cromer and at Malvern with my mother and me and my brother Henry.

———

July 8th.

We visited the Royal Academy Exhibition.

Helen Lambart and Mrs. John Gladstone came to luncheon. Our dinner party:—Lady Alfred Hervey and her daughter, Sir Francis Doyle.

Ethel Methuen (very pretty), and one of her sisters came to luncheon.

We visited Lady Winchelsea, and sat some time with her, she was very agreeable. Willie Bruce and Mrs. John Gladstone dined with us : poor Mimi had a headache.

— — — —

We had the sad news of the death of Lady Lilford (the younger—the wife of Lord Lilford—not his mother). She was an interesting woman, accomplished, very agreeable and very handsome. Her death is to be lamented on many accounts, and especially for the sake of her husband, who himself suffers terribly from the gout, and is especially in need of the loving care of a devoted wife. Lady Lilford is said never to have recovered from the shock given her by the death of her eldest son.

— — — —

We returned to town yesterday from visits which we have been paying to a few of our most intimate friends ; but the pleasure of the time has been a little marred by an accident to my knee. Our first visit on the 10th, was to dear Cissy and Emmy, in their house, "Forest Lodge," on the verge of Windsor Park,—the first time we have seen them in that house ; a very pretty house, arranged and decorated with a great deal of taste and skill. I hardly need say that they were most cordial and kind.

1884.   The next day they took us for a long drive through
the beautiful woods of Windsor ; very enjoyable,
but from want of local knowledge or study afterwards, of the topography, I cannot describe what
we saw. We passed however through the town
of Windsor—through the whole length of the great
avenue or Long Walk—within sight of the glorious
Castle, through a great extent of beautiful woods
(those of Swinley. I believe), and along the side
of Virginia Water.

The woodland scenery is very fine indeed; the old
oaks, beeches, and chestnuts, grand ; some of them
venerable, shattered ruins of vast antiquity, others
in full luxuriance of foliage. The great old elms
in the Long Walk also very grand trees. The
variations of surface, the deep, mysterious looking
dells and recesses, contribute much to the beauty of
the scenery.

The 12th of July, we went by road from Forest
Lodge to Minley Manor, the Charles Hoares' house
—a very pleasant drive for the greater part of
the way through beautiful woods, connected seemingly with those of Windsor; afterwards through the
fir plantations of Sandhurst, and by Blackwater to
Minley. This beautiful place I have described
before.

We were very happy to be there again with dear
Kate Hoare and her husband. But unfortunately
our stay was rather marred by an accident to my
knee, which confined me to bed for a whole day,
and has somewhat crippled me even till now. The
17th however, we went by rail from Farnborough

(the station from Minley) to Weybridge, and thence
in a fly to Fairmile near Cobham, Lady Grey's.
There we spent the next day very pleasantly, and
Lady Grey took us to Thames Ditton to see dear
Annie Campbell and her sister Finetta. The 19th
was a very pleasant day ; we went in a comfortable,
open carriage from Fairmile to Hampton Court ;
there we were most kindly received by Mrs. Ellice
and her daughters, and had a very cheerful and
pleasant luncheon with them and Captain Lambart;
Mrs. Boyle also coming in. Captain Lambart,
(Mrs. Ellice's son-in-law), is a particularly agreeable
man ; Mrs. Ellice herself in excellent preservation,
lively and clear-headed.

I should mention that from Fairmile to Hampton
Court we had for companions in our carriage, two
Miss Sullivans, nieces of Lady Grey's, very nice
girls. From Hampton Court, we had a pleasant
drive to Eaton Place.

————— ——

July 21st.

48, Eaton Place. Stayed at home. Minnie,
Leonora, Katharine, Cissy and Emmy came to see
us; also Fritz Jeune. We took leave of dear Minnie
who is going to Ireland. Edward dined with us.

————— ——

July 22nd.

Montagu, Susan and Car. MacMurdo, and Mimi
Bruce, came to luncheon.

————— ——

July 23rd.

Frank Lyell and William Loch came to luncheon

1884. and dear cousin Catty immediately afterwards; also
    Katharine and Dora.

July 24th.

We went down to Barton, Katharine with us, by
the 2.20 train from St. Pancras. Arrived safe at
our dear home.—Thank God.

July 25th.

Settling ourselves at home; a comfortable talk
with Scott.

A drive in the open carriage with Fanny and
Katharine, and a stroll with them in the garden.
Leonora, Annie Pertz and Rosamond arrived.

July 26th.

Dear Car. MacMurdo arrived, also Mrs. Storrs.

Weather very cold and disagreeable, but I had a
drive with Fanny and Katharine and Leonora.

July 28th.

Dear Katharine, Rosamond and Mrs. Storrs went
away.   Examined flowers of Clerodendron, and
made a note.

July 29th.

Examined flowers of Hedychium, and made a
note thereon.

Had a pleasant drive with Fanny and Leonora
in the open carriage to Ampton and Livermere.

Drove with Fanny and Laura—we visited poor old Patrick Blake.

-- --- ---

August 4th.

Barton. We came down hither on the 24th of July, and are settled now (I hope) for the rest of the year at our dear home. Katharine came with us, Rosamond the next day, and our dear friend Mrs. Storrs, on Saturday ; but these all returned to town on the 28th. Leonora Pertz and her daughter Annie, who came on the 25th, and Car. MacMurdo who came on the 26th, are with us now, and will I hope, remain some time longer.

The weather, these last four days, has been splendid — everything that can be desired for the harvest, which is going on rapidly—indeed with unusual speed—and is very pleasing to see. Mr. Cooper, in this parish, has actually already cut all his wheat, and as the barley is not yet fit to cut, his operations are brought to a standstill for the present.

The famous (or at least much talked about) procession of the Radicals took place while we were still in London ; but I saw nothing of it. On the whole, whatever the numbers, it seems to have been quiet and well behaved. Fritz Jeune, who saw a good deal of it, said he never saw so good-humoured a crowd ; they all looked (he said) as if they were " out for a holiday "—and enjoying it.

--- ---

A really remarkable and uncommon season, a *poet's* August. From the 30th of last month, not a drop of rain; continued sunshine: uninterrupted fine weather, indeed beautiful and very hot. The harvest going on splendidly. We can hardly be sufficiently thankful. Most of the wheat in this parish, I believe, is now cut, and it appears to be in fine condition. The harvest-men, however, have suffered a good deal from the heat.

I see in *The Times* the death of an old friend, General Sir William Codrington—four years older than myself. His father and mother were among the most intimate friends of mine; my brother Hanmer served on board Sir Edward's ship at Navarino, and was wounded there; and as long as we were young the two families continued to be very intimate. Afterwards, as so often happens, without any disagreement, they drifted in different directions —and we comparatively seldom met, especially after my father in great measure gave up London society, and I was a good deal abroad in Brazil, and at the Cape, and I afterwards married and settled in the country.

The last time I remember to have seen William Codrington at Barton, was in 1829, when he came hither with his father and mother, and indeed (I think) all the family, on our return from abroad. The old friendship was renewed between him and my brother Henry at the time of the Crimean war, when the one was in command of the Light Division and the other a Major in the 23rd in that Division;

and from that time they were intimate until Henry's 1884.
death.

Fanny and I, since we have had a house of
our own in London, and have spent there a part of
the spring or summer, have seen Sir William and
Lady Codrington each year. William Codrington
has always been very friendly and cordial to me
whenever we have met. He was not only a very
good soldier, but a very good man; and moreover
one of very pleasant manners.

August 12th.

The wheat harvest is now finished on the Home
Farm here,—on Scott's farm, on Cooper's farm, and
nearly, if not quite, on Baldwin's. The barley is
later.

The weather continues brilliant, hot and dry.
Yesterday, in Scott's garden, at noon, the thermometer in the shade was at 92 deg. Fahrenheit:
and on the 9th at 90 deg. The long-continued
drought has been destructive to the turnips and
other "root crops."

August 20th.

Lady Hoste with Lady Molesworth came to tea.

August 22nd.

Mr. Cooper told me that he has finished his
harvest, and that it has occupied only three weeks

1884. and three days. He remembers only one year (I
think he said), in which the harvest was over in a
shorter time. The wheat crop, I hear, is not very
great in quantity, but most excellent in quality; the
barley very various. The extraordinary long
duration of brilliantly fine weather, hot and dry,
begins to threaten serious consequences : a scarcity
of food for the cattle, and even of water.

We received, yesterday, the news of the sudden
death of Sir Willoughby Jones ; and to-day we have
had the still more melancholy intelligence of Mr.
John Gladstone's departure from this life. I knew Sir
Willoughby but little, but I believe he was a very
good and very useful man, well deserving to be
regretted. His eldest son, Lawrence Jones, has
married Evelyn Bevan, the daughter of our neighbour Mr. Johnson Bevan : and since the marriage
we have seen the young couple often. Evelyn is
one of my favourites among our young lady neighbours, and her husband is intelligent, agreeable,
and I should think well informed.

With John Gladstone, also, my acquaintance has
begun since his marriage. In my journal of April
14th, '83, I noted my very favourable impression of
him, from a visit which he and his bride had paid
us at that time, and I hoped to have many further
opportunities of cultivating his acquaintance. I
little thought that his thread of life would so soon
be cut short ; that he—a young man—would be
called away, while I, more than old enough for his
father, should be left. He has died of consump-
tion, having been ill, I believe, some time. I am

very very sorry for his young wife, a most amiable 1884
and excellent person.

———————

Sir Willoughby Jones' death seems to have been
very sudden; on the Thursday morning (the 21st),
his son, who was at Bury, received from him a
letter written quite cheerfully, and proposing
arrangements for the 1st of September; before
Lawrence Jones was dressed, he received a telegram
announcing his father's dangerous illness, and almost
immediately afterwards came the announcement of
his death.

———————

Lady Bristol sent me a flower of Disa grandiflora.

———————

A fine day—less cold.
We drove to Mildenhall with Willie and Mimi—
inspected the house, &c., and saw the Livingstones:
had a pleasant drive and luncheon. Harry and
Laura arrived.

———————

Pleasant company in our house ever since we
returned from London; first, Leonora and her two
very clever and interesting girls (they left us
September 1st): afterwards, Lina Bruce, William
and Emily (Willie and Mimi) Bruce with their
baby, Susan MacMurdo and dear Car., Mr. and

1884. Mrs. Sidgwick, Lady Alfred Hervey and her daughter Mary, Finetta Campbell, Lily Bruce, Edward; these not all at once, but for various terms: besides numerous *garden parties* at our neighbours' houses, and dinner and luncheon parties at our own. The Bruces, husband and wife, have been delightful. Their child is a most amusing little darling.

Car. MacMurdo is in great beauty; Annie Pertz has painted an admirable portrait of her. Lady Alfred Hervey and her daughter both are very agreeable.

[In a small diary, Sir Charles kept, he mentions the following friends we saw during this month of August, which are not mentioned in his longer journal:—Clara, Lady Rayleigh, Hedley Strutt, Mr. and Mrs. John Paley, Mr. and Mrs. Harry Jones, Mr. and Mrs. James and their son and daughter, Lady Bristol and Lady Mary Hervey, Lina Bruce, and Dora Pertz. We had a large garden party at home, and went to garden parties at Nether Hall and at the Barracks at Bury. Sir Charles was reading Brewer's History of Henry VIII., and writing notes on Mildenhall plants. There was scarlet fever in the village, which prevented us from going to Church ; so we had prayers on Sundays at home.—F. J. B.]

The remarkable drought, which has lasted so long, and made farmers and gardeners unhappy, has broken up at last, without any remarkable storm, and we seem likely to have plenty of rain. The change began on August 27th, when the rainfall amounted to 0·42 inches.—

On the 1st September   . . . 0·29.
,,   2nd   ,,   . . . 0·20.
,,   3rd   ,,   . . . 0·44.
,,   4th   ,,   . . . 0·63.

And in the last 24 hours it has amounted to 1·05.

These last rains have produced a very rapid change in the appearance of our park and lawns; the grass, which had been parched up into almost a whity-brown colour, has now got a cheerful green; and I am told that the rain has come in time to save much of the root-crops.

———

September 9th.

Walked with Laura and Mrs. Horton to "The Cottage," and through it.

Mrs. Mills arrived.

Mrs. Wilson and Agnes, Dorothy Hoste, Mr. Greene, Major Kingscott, Mr. and Mrs. Algernon Bevan, dined with us.

———

September 10th.

Dear Cissy and Emily arrived.

Drove with Fanny and Mrs. Mills.

The Loraines went away.

———

I see in *The Times* the death of our dear old friend George Bentham, at the age of 83 (? 84). He had fallen into such bad health within this last year, and was so much broken down, that one could hardly wish his life to be prolonged, especially as it was very lonely. Some time in last year he was dangerously ill, so that his death was almost daily expected ; but he must have rallied to a certain degree, for when we went to London last spring, I found, on enquiry at his house, that he had gone down to his brother-in-law's, in Herefordshire ; and he had not returned from thence when we were last in London, in July. He died, however, it seems, at his house in Wilton Place.

George Bentham was unquestionably one of the greatest botanists of our time. Of his English contemporaries (those of his generation since the death of Robert Brown and Lindley), I should say that Joseph Hooker alone could be ranked as his equal or nearly so ; and of the foreign botanists coeval with him, I hardly know of five who ought to be ranked in the same class. He seems to have begun the study very early in life, and to have pursued it with undeviating zeal and fidelity to near the day of his death. He was brought up, I believe, in France, and certainly (as he himself more than once told me) learned botany there ; all his botanical studies seemed to have been conducted on the principles of the French school—so much so, that he sometimes seemed to be prejudiced against the Linnean method. His writings and his personal example no

doubt contributed much to establish the French 1884.
system in this country. He was a great traveller.
Had travelled, I believe) in nearly every country of
Europe, and made large botanical collections in all ;
was fond of talking of his travels, and told them
very well. He was particularly fond (I think) of
recalling the memories of a botanical tour which
he made, when a young man, in company with Mr.
Arnott, in the Pyrenees. I remember he mentioned
the stock of botanical drying paper which they laid
in for the expedition ; it was almost incredible. He
seems to have always been industrious in collecting and preserving plants, as well as studying them,
and by the donation of his own herbarium and its
establishment at Kew, he has rendered a most
important service to the science.

George Bentham once told me that he was of the
same age as the century—born, that is, just at the
beginning of it. Where he was born I do not know
exactly, but suppose it was either in Russia or
in France. His father was Sir Samuel Bentham,
who was distinguished in the Russian service ; and
Jeremy Bentham was his uncle.

He was a very amiable man, and a very pleasant
one. His transcendant merits as a botanist are, I
fancy, better known abroad than in England.

Bentham appears to me a remarkable sensible,
judicious botanist, very free from any whims or
caprices, or exaggerated views ; a man of immense
industry and accuracy, indefatigable in studying and
comparing books as well as plants living and dried.
He was really, though *soberly*, enthusiastic in the

1884. pursuit of his science, but I never perceived that he
had any unreasonable enthusiasm for any particular
system or method of study.

Bentham's coolness and soberness of character
was remarkably shown in his latter years, in his
reception of the Darwinian theory. I am not aware
that he had publicly expressed any opinion respect-
ing it, till within the last few years ; then, in the
course of his immense studies of species, genera
and families, the subject was gradually forced on his
attention, and he avowed his adhesion, not with the
eager zeal of a sudden convert, but cautiously,
deliberately and temperately.

I think he published in some scientific periodcal
an account of his early expedition to the Pyrenees,
with Mr. Arnott, but I do not know where to look
for it. There are pleasant notes of his visits to
some other parts of Europe, in volume two of Sir.
W. Hooker's "Journal of Botany," and in volumes
five and six of the "London Journal of Botany."

The remarkably fine weather which I noted at the
beginning of this month, had merely a temporary
interruption, and we have again had a spell of
beautiful days. From September 1st to September
9th there was rain every day ; from September 10th
to September 15th (both included) the rainfall was
0·00; again from September 17th to 21st, rainfall 0·00.

A pleasant walk with Laura and Emmy.

William Bruce and Finetta Campbell came from
Mildenhall. Lady Mary and Lord Francis Hervey,
Mrs. John Paley, Mr. Saumarez, Colonel and Mrs.
Deane, *et al*, dined with us.

The Duke of Grafton and Miss St. John came to luncheon, and we went all through the garden and arboretum with them.

Fanny went to the public opening of the Ladies' High School at Bury.

Harry and Laura went away to London to buy furniture for their "Cottage," in which they are to live here. I shall be very glad when they come back. Laura is a truly loveable creature.

_____

September 16th.

A beautiful day. Dear Minnie arrived. I drove with Fanny and Mrs. Mills to Langham, and saw some of the Wilson girls. Examined flowers of a Canna.

_____

September 17th.

A splendid day, very hot. Lounged with Minnie in the shade. Lady Louisa Legge arrived.

_____

September 18th.

We drove with Mrs. Mills to Mildenhall, and spent several hours there very pleasantly with William and Mimi Bruce and their dear little baby. Visited the girls' school.

_____

September 19th.

Read a good deal of the Memoir of Maria Edgeworth.

I find in the *Gardeners' Chronicle*, of this day, a very good article on George Bentham, in which his merits are duly appreciated ; and I learn from it several facts which are new to me. He was born on September 22nd, 1800, at Plymouth, his father, Sir Samuel Bentham, being then Inspector of the Royal Dockyards. From 1814 to 1826, or thereabouts, he lived with his family in the neighbourhood of Montpellier, and there he first studied botany. His first work—A Catalogue of Plants indigenous to the Pyrenees and Bas Languedox—was written in French, and was published in 1826. I have never seen this, but the writer in the *Gardeners' Chronicle* says—and I have no doubt truly—that there may be perceived in it the germs of those views and methods by which his writings were afterwards distinguished. He became Secretary to the Horticultural Society in 1829, and continued so to 1840 ; during which time he determined and described in the Transactions of that Society all the numerous new species introduced by Douglas, Drummond and others. Thus he continually gained experience, and (what was very important), practice in the comparison of dried plants with the living.

———

September 22nd.

Walked in the garden with Fanny. Read part of Mrs. Gilchrist's book on Mary Lamb.

Lady Florence Barnardiston with 3 daughters, Mr. Wood, Laura and Harry arrived.

Fine autumn weather, rather cold.

Scott arrived from his fortnight's sea-side holiday —I very glad to see him.

Had a pleasant walk with Fanny and Minnie. Mr. Greene and Dorothy Hoste and the Paleys dined with us.

— — —

September 24th.

Mr. Gery Cullum, with Mr. Crowe ("Crowe and Cavalcaselle") and several ladies came to luncheon and to see our pictures.

Lady Florence and her daughters went away. Went on with Mary Lamb.

— — —

September 25th.

Had a pleasant walk with Laura. Mr. and Mrs. Storrs came to luncheon. Mr. and Mrs. Montgomerie arrived.

— — —

September 26th.

A beautiful day. Had a pleasant walk with Minnie and Mrs. Montgomerie. Mrs. Wilson and Agnes came to luncheon.

— — —

September 27th.

Canon Church arrived.

— — —

September 29th.

The Montgomeries went away early.

1884.   News of Patrick Blake's death.

Walked with Mr. Church through the arboretum —gave him my Memoir of my Father.

September 30th.

A beautiful day.   Had a pleasant walk with Laura.   Cissy and Emmie went away in morning. Lady Louisa in afternoon.   Mr. James and his son and daughter and Mr. and Mrs. Morton Shaw dined with us.

LETTER.

Barton, Bury St. Edmund's,

October 1st, 1884.

My dear Edward,

I was very glad to receive your letter from Baden Baden this morning, and to find that you had enjoyed such an agreeable tour, and such fine weather. As I did not know where you were, I had written a letter to you yesterday to your London Home to tell you of the sudden death of Patrick Blake, which happened in his own house at Thurston, just as he was about to breakfast.

I cannot consider it otherwise than as a happy event for him—a happy winding up of his course; at his age (86 or 87, I am not sure which) with health so much broken, and such a lonely life, I think his very best friends could not have wished his life to be prolonged.   It has been very apparent the last year or two, how much he was broken, and one hardly ever saw him that he did not ex-

patiate mournfully on the death of his old friends.
Peace be with him ; he was a good warm-hearted
honest-hearted, honourable man, and certainly the
very oddest and most eccentric I have ever known.
A few weeks ago died George Bentham, not much
younger than Patrick Blake—very unlike him in all
else ; but we might say of him also, that he had out-
lived all that could make life desirable, and that for
his own sake one could not wish it to be protracted.
He was certainly one of the greatest botanists we
have had, but I suspect his fame was more widely
spread abroad than in his own country, and he was
a very pleasant man too.

I am very glad to hear that you are well, and have
enjoyed the same extraordinary fine weather as
we have. I am thankful to say that we also have
been as well as usual (except that I had a sharpish
bilious attack about ten days ago, but I soon got
over it). We have as usual been quite stationary, but
have had an abundance of pleasant guests. Harry
and Laura are in excellent spirits and very busy,
superintending the progress of their cottage, which
is still far from completed ; but I hope it will be
habitable before the cold weather sets in.

With much love from Fanny,

Believe me, your affectionate brother,

CHARLES J. F. BUNBURY.

JOURNAL.

1884. A most beautiful morning.

Mr. Church went away.

My Barton rent audit: the tenants friendly, but melancholy.

Fanny, Laura, and Harry, joined them and me at the luncheon.

Patrick Blake died on the 29th of last month, very suddenly, in his own home, as he was about to go to his breakfast. It is an event which cannot be regretted by his best friends. At his age (86 or 87), completely broken as he was in health, very lonely, having survived nearly all his old and particular friends, his departure can only be regarded as a most happy release. He was an extremely odd character; certainly a warm-hearted and honourable man, but decidedly by far the oddest and most eccentric I have ever known.

In the lifetime of our friend Lady Cullum, it used to be very amusing to hear the skirmishes, in words, between her and Patrick Blake. They were great friends, but very much opposed in their views on many subjects; and both having strongly marked characters and habits of speech, the encounters of their wits used to be very diverting.

Patrick could not be called a regularly well educated or well informed man, though having an active mind, he had picked up a good deal of curious knowledge on various subjects in the course of his wanderings about the world. But what often struck one as peculiar — considering his circumstances—was his fondness for Latin literature, and especially for the Latin poets. He had an excellent memory, and in his old age, even to within the last year or two, his mind seemed to cling with singular tenacity to those studies of his boyhood. He often, in these last years, expatiated to us on the delight he found in reading over and over again Horace and Virgil and Ovid.

October 3rd.

A visit from Lord Waveney.

Resumed the reading of Brewer's Early reign of Henry VIII.

October 4th.

The eclipse of the moon.

Mrs. Mills went away. We, with dear Car, went over to Mildenhall and spent the day pleasantly with the dear William Bruces, taking our leave of Mimi and her baby.

A brilliant day, with very cold wind.

October 6th.

Arthur arrived — also Mr. and Mrs. and Miss Walrond, Lord John Hervey, Mr. Douglas Tollemache, Mr. Trollope, Sir Brampton Gurdon, *cum al.*

Disappointment of William Napier and his girls and of Emily Bunbury.

October 7th.

Patience Morewood arrived, also Cecil, *cum plur. al.*

Fanny's ball.   Wrote to Lord Stradbroke on behalf of Harry.

October 8th.

A beautiful day. A drive in pony carriage with Fanny. Read a little more of Brewer. Pleasant talk with Mr. Walrond.

October 9th.

A courteous and pleasant answer from Lord Stradbroke.

Received Nicholl's application for Lord Sandwich's rent—answered it.

William Bruce came to luncheon. The Bury Ball, all went to it except Minnie, Laura, Walrond and myself.

====

LETTER.

Barton Hall, Bury St. Edmund's,
October 9th, 1884.

My dear Katharine,

Very many thanks for your pleasant letter, which I was very glad to receive.   I have been long intending to write to you, but between company, laziness, constitutional exercise and a little study, I have always put it off.

I am much interested by your remarks on dear

old Bentham's death, and agree with you entirely.
One can hardly tell whether one ought, or ought not
to feel regret for his departure. For himself—he
seemed so completely broken in health, incapable
of his usual occupations, and living such a lonely
life—that I think we can only regard it as a
happy release. Life could offer him no more enjoyment ; and he had done his life's work, and done
it admirably. But to the few friends who saw him
constantly to the last, it must be a loss.

The article in *Nature* was sent to me by an
unknown friend, and interested me extremely. It
must certainly be by Joseph Hooker; the total
omission of *his* name would have been incomprehen-
sible and inexcusable if it had been anyone else. I
was not only interested, but learned a great many
facts from it about Bentham and his family and
friends. I had no notion that his life had been so
varied and full of incidents (though I knew that he
had been a great traveller), nor was I aware that he
had ever tried to practice as a lawyer. I think he
was very wise in giving up that scheme entirely, and
devoting himself to botany. I cannot imagine him
as a lawyer.

I quite agree with you, that (as far at least as I
know of the botanists of the present day), Joseph
Hooker and Dr. Asa Gray, are the only ones
worthy to be placed in the same rank with Bentham,
but Mr. Baker (who worked up the lilies so well),
Mr. Ball (of Morocco), and perhaps Dr. Maxwell
Masters, are quite worthy I think, to step into the
places which the front rank may leave vacant.

1884.   What you say about Patrick Blake is both just and kind. One cannot help feeling some regret for so old a friend and so good-hearted a man ; but he was so altered latterly, so broken, so lonely, and felt so much the passing away of so many of his old friends, that I really felt one ought to rejoice rather than sorrow for his release.

(Oct. 10th).— We have had a glorious, a really wonderful summer and autumn, and have enjoyed them heartily, and I am very glad you have had the like enjoyment in Scotland. This morning, at last, it has been raining heavily, and looks like a continuance of the same ; this will be welcome, for there is already some real inconvenience, and apprehension of more, from scarcity of water. Yet the grass still looks green and fresh — owing, I suppose, to the heavy dews at night. The autumnal colours of the foliage are now very beautiful. I do not think I ever saw them more so. In particular, the Mespilus Canadensis, a small tree, standing opposite to the S.W. end of the drawing room (do you remember it ?) is a mass of foliage of the brightest crimson. In the hedges too, the profusion of fruit on the hawthorns, wild roses and brambles, is something quite remarkable.

We have had the house cram full, this week, of guests for the Bury ball, and some very pleasant people. William Napier, indeed, failed us by reason of illness in his family ; but we have had Mr. Walrond (whom I like particularly), Lord John Hervey, and a number of young *dancing* ladies and gentlemen.

The William Bruces leave Mildenhall to-day for London; I am sorry for it; it has been a great pleasure to have them within our reach; and we went over twice to spend the day with them; they are a delightful couple.

Harry and Laura are very busy superintending the repairs and alterations of " The Cottage," which (as always happens) take up a great deal more time than was reckoned upon. They are as happy a couple as one could wish to see, and it is very pleasant to have them here. I grow more and more fond of Laura.

The last time we heard from Edward, he was at Baden-Baden, and was about to proceed to Munich —after which his plans " depend almost entirely " upon the weather."

Fanny, I am thankful to say, appears to be tolerably well, and not as yet the worse for the *racket* in which we have lived for several days past. I am quite well. Give my love to Rosamond, and the rest of your family, and

Believe me ever,

Your loving brother,

CHARLES J. F. BUNBURY.

JOURNAL.

October 11th.

The gay party, of mostly young people, which had assembled here for the Bury Ball, is to day dispersing. They have been very merry, and the

1884. girls were a very pretty bevy—Car MacMurdo and Dora Walrond conspicuous for beauty. Patience Morewood may well be reckoned among the *beauties*, but she is not a *girl*.

We were disappointed of William Napier, who was to have been of the party. Walrond was most agreeable, and I had a great deal of interesting and satisfactory talk with him. I am particularly fond of his company : he has so much knowledge both of books and men, with such a modest, quiet, pleasant manner, and with little touches of quiet humour. Lord John Hervey also was very pleasant.

Our dear William and Mimi Bruce, after staying here with us from August 12th to September 3rd, removed on the latter day to our house at Milden-hall, which we lent to them ; they remained there till yesterday, and we visited them twice from hence. They are always delightful. Their baby is not only a fine child, but very amusing. They are now gone to London for her "confinement."

————

October 12th.

Before the Bury Ball party, we had had several other pleasant guests :—our old friend Mrs. Mills (who stayed with us nearly a month), Lady Louisa Legge (who is very entertaining), Mr. Church (brother of the Dean of St. Pauls, a very well read, earnest, thoughtful, interesting man, of mild and pleasing manners),—*cum multis aliis.* In fact, since we came from London, we have hardly ever been alone. It is a real privilege to converse with such men as Mr. Church and Mr. Walrond.

Our summer and autumn have been splendid— 1844.
really wonderful; certainly since 1868 or 1869 we
have had no such seasons with such long and
almost uninterrupted duration of sunshine : for even
in these latter days, though the wind had turned
cold, the sun was bright and of some power. At
last the change, long expected, has taken place ; on
the 10th there was heavy and long continued rain,
with a violent wind, and the thermometer in the
night went down to 34 deg. Yesterday was very
cold and stormy, and to-day seems to be the same.
In the 24 hours, from 9 a.m. of the 9th to the same
of the 10th, the rainfall was 0·27 inches : in the
10th to the 11th it was 0·80 : in the next 24 hours
0·34 : and the weather does not look at all settled.

The autumnal colouring of the trees and shrubs
has been, and indeed still is, very beautiful ; the
Mespilus Canadensis gorgeous in scarlet or crimson : the Liquidambar exquisitely variegated with
many shades of red and green : the Horse-chestnuts
(many of them) splendidly yellow. There has been
no decided frost as yet, and the tender plants in
the flower-beds are as yet untouched,—even the
Castor-oil plant and the Fuchsias. But since the
19th of September (which was a very cold day,
though not wet), I have only once seen a butterfly
in the garden (except the common Cabbage White);
whereas before that, the flowers of the long bed
were frequented by plenty of Scarlet Admirables,
Tortoise-shells and Painted Ladies.

Our Vicar, Mr. Harry Jones, came to luncheon
with us—quite freshly returned from his visit to
Canada, whither he went with the British Association
to its meeting at Montreal. He seems in excellent
health and spirits, and to have greatly enjoyed his
expedition. He did not merely go to Montreal,
but was there invited to join the select party of
scientific men which went on to the Rocky Mountains; accordingly he proceeded to the actual ridge
or water-shed of those mountains, even a little
beyond where the railway at present ends. He
says that the country which he traversed in this
immense railway journey, from the Atlantic to the
Rocky Mountains, may be classified under three
great regions :—

1.—The region of forests (coniferous trees pre-
dominating), from the Atlantic to beyond the great
Lakes.

2.—The Prairies—the bare, open, treeless country
to the foot of the Rocky Mountains.

3.—The Rocky Mountains themselves, of which
(in that part at least) the scenery is quite in the
Swiss style.

The railway of which he spoke is, I believe, the
Canadian Pacific.

Mr. Jones's talk was full of matter, but he spoke
so fast that I lost much of what he said.

He said that the way in which the scientific party
was received and treated, was really magnificent;
all their expenses were paid, and their journey
seemed almost like a Royal progress.

Nothing (he said) could be more gratifying than 1884.
the feeling shown towards England throughout the
Dominion.

————

Arthur started on his return to Dublin.   Wrote a
note on Nicotiana aff.   Read some more of Brewer's
Henry V.III., and wrote a little of my Reminiscences.

————

Dear Leonora and Annie arrived.

Read the *Edinburgh Review* on Lord Malmesbury's
Memoir.

————

The mild—even warm—weather continuing.

Read *Edinburgh Review* on the Merivales.

John Herbert arrived.

Some arrangement in my herbarium.

————

We called on Mrs. H. Jones, and saw some photo-
graphs of Canadian scenery.

Finished my list of grasses in my herbarium.

Went on with my Reminiscences.

————

We went to morning Church, and received the
Communion ; heard an excellent sermon from
Mr. Harry Jones.

A beautiful day.

Walked with Leonora through the arboretum, &c.

Began to read Memoir of Admiral Markham.

Went on with my Reminiscences.

John Herbert went away, also Clement.

———

October 23rd.

Dear Leonora and Annie went away. I was very sorry to part with them.

Lord Euston came to luncheon.

Went on with my Reminiscences—also reading Memoir of Admiral Markham—also studying Melastomaceæ.

———

LETTER.

Barton,
October 24th, '84.

My dear Edward,

Many thanks for your letter, which I was very glad to receive, as I was beginning to wonder what had become of you, as it was just three weeks since we had last heard from you; and though we did not suppose that you had met with any adventures, we could not know that you might not be ill. I am very glad that you are and have been well, and have made a pleasant tour, though among places familiar to you, and that you have had fine weather to enjoy it. As for that, the weather can hardly have been finer anywhere than it has here. It is really an astonishing season. I can

hardly remember such a long continuance of bril- 1884.
liantly fine weather, and that following a hot and
dry summer. To-day, when Laura and I were
taking a walk in the middle of the day, the sky was
almost absolutely cloudless, and the sun quite
powerful; and so it has been for many days. In
several places I hear an inconvenient scarcity of
water is felt. The autumnal colouring of the leaves
is most beautiful. I never saw it finer. Fanny
desires me to say, with her love, that we should
be very glad to see you, but she must *count beds*
before sending a definite invitation, and she will
write to you on Sunday next.

Believe me ever,
Your affectionate brother,
CHARLES J. F. BUNBURY.

JOURNAL.

October 25th.

Morning very fine with slight frost—afternoon
foggy.

Consultation with Fanny and Scott about Mildenhall Workhouse.

Went on with my Reminiscences; read some
more of Admiral Markham.

October 27th.

Leopold and Lady Mary Powys, with two daughters, Mr. and Mrs. Sancroft Holmes, Captain
Markham, and Nelly Wilson arrived.

1844.   Went on with Admiral Markham—wrote some more of my Reminiscences, and studied Triana's Melastomaceæ.

<hr/>

October 28th.

Rough weather—wind and rain.

Received a letter from Mr. Livingstone about the Mildenhall Union Workhouse—answered it—much talk on same subject with Fanny and Scott.

Fred Freeman and Mr. Wingfield arrived. The *fancy ball* at Bury.

<hr/>

October 29th.

A beautiful day with slight frost in morning.

Had a pleasant walk with Mrs. Holmes. Studied Triana's Melastomaceæ—went on with Brewer's Henry VIII. The *fancy dress* dance in the breakfast room—*amabilis insania.*

<hr/>

October 30th.

Scott's report of the meeting at Mildenhall, about the workhouse—important.

Lady Florence Barnardiston and her daughter went away.

Finished reading the Memoir of Admiral Markham.

<hr/>

October 31st.

A beautiful day.

A pleasant walk with Minnie and Mrs. Holmes.

The Leopold Powyses went away ; also Captain Markham.

Went on reading Brewer's Reign of Henry VIII. 1884.
The weather of this autumn has continued to be
very extraordinary and exceptional — not, indeed,
uniformly mild, but with a strange absence of rain,
and yet almost entirely without frost. In this month
there have been eighteen days without any rain, and
five other days with only ·01 or ·02 of rain. There
were, however, a few days, chiefly about the
middle of the month, with heavy rain ; 0·27 on
the 10th.

The garden has been in great beauty, and even
now there are still flowers enough, especially in the
long bed, to make a handsome show. Though there
has been frost once or twice at night the Castor-oil
plant, the Fuchsias, the Dhalias, the Zinnias show
no sign of it, and do not appear to be at all hurt ;
nor do the Cannas, as to their leaves.

The autumnal colouring of the foliage has been,
and indeed still is, uncommonly beautiful ; I do not
remember to have ever seen it more so. Many of
the exotic trees, indeed (the Canadian Medlar for
example) have now entirely lost their leaves ; but
some, in particular, the Liquidambar and the Acer
rubrum, are in glorious beauty of colouring, in many
various shades of red and yellow. The large trees
have for the most part, turned more or less completely yellow (the Beeches golden brown) but the
common Oaks are very various, many keeping
unchanged their deep green, while others have
turned very yellow.

Much gaiety and lively society through nearly the whole of last month ; a predominance of young people ; two balls at Bury, one of them a *fancy* one; and two dances in this house, one of these also in *fancy* dress. This last, really rather a pretty and amusing affair, and the young people seemed to enjoy them all.

We have hardly ever been alone (and I mean to include in *alone* the times when we have Minnie and Laura and Harry with us). We had, however, a quiet, comfortable, rational breathing time from the 19th to the 26th ; and, as I noted before, it is a real privilege to converse with such men as Canon Church and Mr. Walrond. Latterly too, we have had some agreeable guests — Leopold Powys, Captain Markham, Fred Freeman, Mrs. Sancroft Holmes.

Dear Car MacMurdo left us this morning, to join her father and mother who are going to Alassio. I was very sorry to part with her. She has been as charming and loveable this year as last, though, as she has not been alone with us, I have perhaps not felt her merits as strongly.

I fully confirm and repeat all I have said about Laura's merits ; she is loveable and charming, clever and good, and it is delightful to see her and her husband so happy.

———

November 2nd.

Acer rubrum is now more splendid — more glowingly beautiful, than I have yet seen it.

A very fine day, very mild.

A long discussion with Fanny and Scott about Mildenhall Workhouse.

Read 1. Samuel, chapter 1. Read an ode of Horace, lib. 3.

Death of Mr. Fawcett, Post-master General.

We have had the good news of dear Mimi (Emily) Bruce's safe delivery and the birth of her little girl, her second child. God grant her a good recovery.

We have lately had a visit, though a very short one, from our dear Bishop of Bath and Wells; he came to us with his usual kindness, to preside at a meeting, at Bury, of the Suffolk Branch of the Society for Prevention of Cruelty to Animals, Fanny having earnestly begged him to give it his support.

The meeting held in the Guildhall at Bury, on the 5th, was a very good one—a decided success. Lord Arthur spoke excellently well, as he always does. Mr. Rodwell, Archdeacon Chapman, Mr. Abraham, also very well; the Duke of Grafton well, far better than his father or his uncle, Lord Charles, ever did. I hope it will do good.

We went in great force, having invited a large party on purpose. Beside those I have just mentioned, we had Admiral Spencer, Mr. and Mrs. Clements Markham, Mrs. Wood (Laura's mother) and her daughter Mary. The Bishop of Bath and Wells is looking remarkably well, seems in excellent spirits, and is delightful as ever. He is certainly the youngest man of his age that I know ; he was 76

1884. last August, and I am sure he does not look much above 60. Long may he flourish.

Mary Wood was, I think, the only one of the party who had not been here before; she is an agreeable young woman, lively, clever and well informed, with a very sweet voice.

News of the death of Mr. Fawcett, the Post-master General:—a great public loss. An especial loss he must be to the Ministry, of which he was one of the best members; indeed he was in my opinion, one of the best and wisest of all the Radicals.

November 10th.

Laid up all day (confined to two rooms) by a bad cold caught in Church on Sunday. Read a good deal of Lord Malmesbury.

November 11th.

Better, but confined to my two rooms.

Read much of Lord Malmesbury, and of Sydney Smith's Memoirs.

Dear Sarah Seymour arrived.

November 12th.

Still a prisoner. Visited by William Napier as well as by Laura.

November 13th.

Finished Vol. 1. of Lord Malmesbury, and read much of Sir George Napier's Life.

Talk with Edward.

Had range of three rooms on 1st floor.

Went on with Lord Malmesbury and with Sir George Napier.

Pleasant talk with Laura and with Albert and Sarah.

November 15th.

Harry and Laura went away on some visits.

Went down to library in evening—pleasant chat with William, Edward, Minnie, Fanny and Miss Wilmot.

Finished Memoir of Sir G. Napier.

November 17th.

All our guests went away except dear Minnie.

Finished Horace Walpole and his World, and went on with Lord Malmesbury.

November 21st.

Wrote a little of my Reminiscences. Finished Stuart Reid's Sydney Smith. Went on with Lord Malmesbury.

Went down to library in evening, and dined there with Fanny and Minnie.

November 22nd.

Down to my old study in morning, and stayed there above two hours with great comfort and satisfaction. Finished Lord Malmesbury.

A sharp white frost.

Spent the middle part of day pleasantly in my old study.

Read part of St. Matthew's Gospel, chapter 8, in Greek. Re-read part of Greene's History of England and of Hallam's Lit. Eur.

LETTER.

Barton Hall, Bury St. Edmund's,
November 27th, 1884.

My dear Katharine,

I saw in *The Times*, yesterday, the news that Godwin Austen had "gone before" us; and the news gave me a solemn feeling as my memory travelled back over the many years during which I had been used to meet him at the Geological and at Charles Lyell's and your father's. One of the last times that I met him, I remember he said to me, that he no longer cared to go to the G. S. Meetings, for the "old set" was all gone; and I quite agreed with him. He will have left a name in geology which will not soon be forgotten. He was a good and able man, and a pleasant one too, though "he had his fancies." I am very sorry for Mrs. Austen.

And now I see, today, the departure of Bonham Carter:—I knew him much less intimately than I did Austen: but I fancy that he and his family were old friends of yours, therefore I am sorry.

Truly, the *warnings* come fast when one comes to my time of life.

Thank you much for your kind and pleasant letter of the 21st. My cough has been very tiresome and annoying, but I hope it is now gradually departing, though very slowly : and certainly it is impossible for any one to be better taken care of than I am. I am very sorry that dear Rosamond suffers so much from neuralgia.

I am very much interested by your account of your visit to Mr. Arnold's school at Eversley, and it is delightful to hear that dear Charlie is established in such a charming place. I, like you, delight in that heath and fir style of country. I have a dried specimen of the Marsh Gentian (Pneumonanthe) from near Eversley, given me by Mrs. John Martineau, but it is diminutive to what I have gathered in Switzerland.

Having been confined to the house now for very near three weeks, I have had plenty of time for reading, but I have not got through any very important work. I have been a good deal entertained by Lord Malmesbury, particularly the 1st volume. Such a rapid survey of our life-time (I say *our* because *his* has nearly coincided in time with mine), shows in a striking way what a multitude of extraordinary events and changes have happened in that time. One can hardly believe that we have continued so little changed (at least so it seems to *me*), while such wonderful changes have been happening all round us. The new Life of Sydney Smith, also, I thought well and pleasantly written,

1884. but I do not know whether there is much in it
that is new.   It is long since I read the Life of
the same Sydney by his daughter, but my impression is that it was very well done.

It will be three weeks tomorrow since I have set
foot either in my museum or my garden !—and in
that time the season has completely changed : this
day there has been a regular fall of snow !

I am very glad that Joseph Hooker has given
you an interesting relic of Bentham.

With Fanny's best love.

Believe me ever,

Your loving brother,

CHARLES J. F. BUNBURY.

JOURNAL.

November 29th.

Snow !

I convalescent.

Read St. Matthew, chapter 9.

Read some of Le Misanthrope, and re-read part
of Life of Sydney Smith by his daughter.

December 1st.

Sarah and Albert, Laura and Harry, Annie
Campbell and her brother Gerald, May Frere,
Mr. and Mrs. Livingstone, Mr. and Mrs. Barber
arrived.   Mr. Bevan and his daughter dined
with us.

The Livingstones and Barbers went away early.
Colonel and Mrs. Moncrieff arrived.

— — — —

December 3rd.

Mrs. Wilson came to luncheon.
The Loraines and the Blands dined with us.

— — — —

December 4th.

A very large dinner party:— the Hostes, Lady
Rayleigh, the John Paleys; *cum plur. al.*

Wrote a little of my Journal.

Read some of Green, Hallam's Lit. Europe, and
Memoir of Sydney Smith.

I have been confined to the house for nearly a
month—ever since the 10th of November—by a
cold and cough caught in Church on the 9th ;—am
now nearly well, but not yet able to go out; this
however I do not care for. I have been most delightfully nursed by my incomparable wife. Most
part of the time we have been nearly alone, having
no guests except dear Minnie; but she is a great
comforter. Now again we have a house full of
company.

During the time that I have been a prisoner,
there has been a complete change of weather—
indeed of seasons. The beautiful, mild, almost
summer-like weather we had enjoyed so long, lasted
till near the middle of November, and the trees
retained a great deal of their variegated beauty
of colouring ; but not long after, rough and stormy

1884. weather set in, and the groves were rapidly stripped of their beauty. But *snow* was still unexpected, when it began on the 29th.

On the last day of November, there was a really heavy fall of snow, and it remained, thickly covering the ground all through the first day of this month. A dismal fog and thaw set in on the 2nd ; mild weather on the 3rd, all the snow having disappeared. except from shady nooks.

December 6th.

All our company departed, except May Frere and (of course) Laura and Harry and Minnie.

December 8th.

May Frere went away.

The dear little Seymour boys, Charlie and Bill, arrived.

Began reading the Croker Papers.

December 9th.

We had the great pleasure of a visit from dear Sarah and Albert Seymour, who stayed with us from the 1st of this month to yesterday :—pleasant as ever.

We had also Colonel and Mrs Moncrieff, Annie Campbell and her brother Gerald, May Frere, and Laura and Harry (just now returned from London); —a pleasant party. The Moncrieffs, both husband and wife, are very agreeable and clever.

Read a little of Cicero de Senectute. Went on
with the Croker Papers and with my Reminiscences.

————

December 13th.

A fine day and mild.

Dear Charlie Seymour went away.

Read Cicero de Senectute. The Croker Papers
as before.

————

December 15th.

The end of the year is near at hand, and again I
feel myself (as so often before) called upon both by
duty and inclination, to offer my most humble and
earnest thanks to Almighty God, for the innumer-
able blessings bestowed on me, and of which I feel
myself very unworthy. Above all, I am grateful,
and never can be sufficiently grateful, that my
admirable wife is preserved to me, and that we live
together in uninterrupted harmony, and in the
enjoyment of good health, considering our ages.
Surely no man ever could be more fortunate in a
wife than I am ; I can, indeed, never be thankful
enough for such a blessing.

I hurt my knee by an accident while we were
staying with the Charles Hoares at Minley, in July,
and was in some degree crippled by the accident for
some weeks. And lately, as I noted on the 4th of
this month, a rather severe cold kept me a prisoner
for about four weeks : but neither of these incon-

1844. veniences can be considered as serious drawbacks
from my general good health ; for which (my health)
I have great reason to be very thankful.

We have lost, in this year, several old and good
friends, besides others with whom our intimacy had
been less. George Bentham, indeed, and Patrick
Blake, could hardly be regretted, for both were so
completely broken down by age and infirmity, and
the lonely position of each was so melancholy, that
we could hardly wish their lives to be prolonged.
Lord Arran also was very old, and we had of late
seen him very seldom, so that regret must be chiefly
on account of his wife and family. Lord Hertford
was not young, but he seemed so full of life and
spirit, so happy himself, and diffusing so much
happiness around him, that I really grieved for and
with his family. I might say something of the same
sort of Sir Bartle Frere.

Godwin Austen we have known a long time, and
for several years had been almost intimate with him,
when I used to meet him constantly at the Geological Society meetings, and often at Lyell's and Mr.
Horner's. Mrs. Austen also, was a particular friend
of Susan Horner, through whom we used to hear
very often of them ; but though we have stayed at
both their homes—Chilworth and Shalford, I never
felt nearly so well acquainted with her as with her
husband. Austen was a really eminent geologist—
zealous, industrious, acute ; a good observer, and a
good describer of what he observed ; bold in his
speculations, and perhaps sometimes a little fanciful.
He was a good man too, and I believe, very useful

as a country gentleman, after he came into his 1884.
property at Shalford.

I do not know how it happened that in this obituary I forgot Sir William Codrington, a very old friend, though one with whom we have not of late years been very intimate.

I might also have mentioned Lady Herschel ; and certainly ought to have omitted Lady Lilford ; who died on the 10th of July, and who was a person much to be regretted.

Other acquaintances who have passed away in the course of this year are :—John Gladstone, whom we had known but a short time, but to whom we had taken a great liking ; Milner Gibson, with whom I have been acquainted many years, but never intimate.

John Bonham Carter, with whom also I never was at all intimate, though he and his family were old friends of Fanny's people.

Truly, we are " like stranded wrecks " waiting "on the verge of dark eternity."

My deafness is a great inconvenience to me, a great hindrance to my enjoyment of Society, but I cannot distinctly perceive that it has increased since last year. The decline of my memory (especially as to proper names) is likewise inconvenient, but neither in this can I clearly perceive any aggravation.

———

December 16th.

A fine day.

Received a letter from Mr. Charles Kraus, from

1884. Pardubitz, Bohemia, asking for a copy of my " Botanical Fragments." Sent him one.

December 18th.

Violent wind and rain—excessively cold.

Paid Dr. Kilner for medical attendance on poor.

Studied a little of Geikie's Geology.

I noted in this Journal (March 11th) the marriage of my nephew Harry to Laura Wood, and have repeatedly mentioned her since. I need only add that the marriage has fulfilled all the best hopes that we could have formed of it. Laura is a truly delightful and loveable creature, and contributes greatly to our happiness and to the cheerfulness of our home. She and her husband are truly devoted to each other, so that it is a comfort to see them. All that we can wish and pray for, is that they may long be preserved to each other.

My nephew George Edward has also been married in the course of this year, in Canada.

December 20th.

We have this year again had the enjoyment of much agreeable society, some of which I have mentioned in this volume.

Our dear Minnie Napier, who is quite as a sister to us, left us this morning, having stayed with us from the middle of September ; she is always pleasant, and her company was particularly cheering while I was so long imprisoned by my cough. We had, not long ago, a very pleasant though short visit 1884. from dear Sarah and Albert; and have since had their two dear little boys staying with us.—Charlie, however, only for a week.

Since the beginning of this month, there has been much rain, which must be considered as an advantage, after such a remarkably long continuance of dry weather;—the last winter having been uncommonly dry, as well as the spring, summer, and autumn. The rain must be considered as a blessing, specially because of its cleansing and disinfecting properties, its effects in promoting the action of drains and helping to remove dirt. Its help in these ways seems to have been particularly needed in the case of the Thames, which according to the newspapers, had been brought into a shocking state by the drought.

The harvest this year has been good, though not very remarkably so; but the excessively low prices of all agricultural produce, especially of wheat, have been very alarming and distressing to farmers. Indeed the prospect for us who depend on agriculture is decidedly unpleasant; for there is no probability (as far as I can see) that the prices of farm-produce will become in any important degree higher; rather it may be likely that they may fall still lower; and if so, how will it be for possible for farmers to meet the many and heavy expenses of taxation, labourers' wages, &c. ? The burden must ultimately fall on our (landlords) rents, and it may do so very soon. The prophecies or forebodings of the Protectionists seem to be coming

1884. true, though they have been much longer about it than Croker and the leaders of the party expected.

LETTER.

Barton Hall, Bury St. Edmund's.
December 22nd, '84.

My dear Edward,

I wish you a merry Christmas and a happy New Year.

I caught cold in Church on the 9th November, and a troublesome cough hung about me so obstinately that I did not get out again till the 7th of this month. I am now however quite well, and I am thankful to say that Fanny is well likewise. We are now alone with Harry and Laura, for Minnie left us the day before yesterday, and we shall have no Christmas party; but we expect some company for the Bury Ball about the middle of next month. Harry and Laura seem very happy, and are very busy about "the Cottage," which is not yet habitable, and required a great deal more time to make it so than was expected when they first came.

Since you were here, two old friends have departed this life :—Godwin Austen and Bonham Carter; with the latter indeed, I was not intimate as I believe you were, but his family and connexions were so intimate with Fanny's people, that I *seemed* to know much about them. How our contemporaries are dropping off!—going before us.

While I was confined to the house by my cough, 1884.
I read through Lord Malmesbury's two volumes,
which entertained me much, especially the first,—
and also through two volumes of Croker, in which I
found a good deal to interest me, especially in
what the Duke of Wellington told; but I confess
I am now getting a little tired of Mr. Croker and
his politics.

Again I will wish you very heartily a merry
Christmas and a happy New Year, and as many of
them as may be conducive to your happiness, and
with Fanny's best love, believe me,

Your affectionate brother,

CHARLES J. F. BUNBURY.

JOURNAL.

December 23rd.

Wrote to Mr. Livingstone about the Mildenhall
almshouses.

December 25th.

Read prayers with Fanny—read also ch. ii. of St.
Matthew. and ch. i. of St. Luke, in Greek.

December 26th.

Mrs. Wilson and two of her daughters came to
luncheon—pleasant.

December 27th.

Very dark and very cold.   Read Cicero de Senectute.

Wrote to Minnie.

I will say little about politics for it is a subject (I mean the politics of our own day) which is disagreeable to me, and on which I do not willingly allow my thoughts to dwell.   The only incident connected with such matters, which has lately, or for a long time past, given me any satisfaction, is the recent compromise between the two houses on the subject of the franchise—the chiefs of the two parties consulting together, and trying to find some means of reconciling the conflicting views.   This was a wise and good step.

I dislike indeed the extension of the elective franchise to rural households ; I am afraid it will be productive of mischief (though not immediately) —but as it seems to have been thought hopeless to contest that point, the step taken by the Conservatives seems the wisest of which the state of the case admitted.

The Soudan war drags its slow length along, and as far as I can see, the best we can hope for is that it may not end in some terrible catastrophe.

The earthquake, which was so severely felt in Essex last April, was a very remarkable event, for I do not remember that one so severe is recorded to have happened in England—certainly not in modern times.

The sudden death of the Duke of Albany (Prince Leopold) on the 28th March, excited a general

feeling of sorrow.   The sentiment of attachment to
the Queen and her family is, I hope and believe,
still strong and wide-spread.

The villainous attempts to do indefinite mischief
by means of explosions, have been repeated more
than once in this year ; but the Almighty has
mercifully preserved us from any extensive results
from the wickedness of such villains.

I have not followed up my Botanical Fragments
by any further attempt at writing anything on the
same subject which might be printed.   It is
perhaps a pity that I have not done so ; I have
now and then felt a little the want of some object
for writing.

In the early spring I wrote a little sketch of the
botany of this parish, at the suggestion of Mr.
Harry Jones, who had a scheme of drawing up
something in the way of a Magazine or Almanack
of this parish, and asked me to contribute.   I
wrote my contribution and sent it to him : but I
have heard nothing more of his scheme.

I have also completed as far as I could, my list
of wild plants of Mildenhall parish—as far as I
could, because, while we were living at Mildenhall,
various circumstances, not now to be recorded,
interfered with my attention to botany.

[Both these papers have been printed by me since
Sir Charles's death, in a volume called "Botanical
Notes."—F. J. B.]

1884.

Read Cicero De Senectute.
Wrote to Lady Muriel Boyle.

* * *

A beautiful, still, sunny day, with slight frost in
morning.   Walked with Laura.   Mr. and Mrs. and
Miss Lushington came to luncheon.

* * *

Finis Anni
1884.
Laudes Deo.

1885.

JOURNAL.

A very cold, dark, disagreeable day.   Walked
with dear Laura.   Read some of Croker.

* * *

Read Cicero de Senectute.   Read, in *Nineteenth
Century*, Max Müller on Savages.

* * *

A sad and terrible disaster has befallen some of
our neighbours.   Ampton Hall was burnt to the
ground early yesterday morning : it is a mercy that

no lives were lost, nor anyone seriously hurt, and 1885. that, as the fire begun near the top of the house, it was possible to save many articles—the plate, Mrs. Paley's jewels, many pictures and a great many books. Arthur, who went over immediately after breakfast to see the real state of the case, says that the servants' wing of the house is nearly uninjured, but the main building is reduced to a mere shell. It is a grievous misfortune, and I am very sorry for the Paleys. The fire seems to have been occasioned (as is so often the case), by an overheated flue for hot air, which set fire to a beam in a "box room" in the upper part of the house. It was first discovered by a maid who (fortunately) was kept awake by ill-health, and smelling fire, roused the family.

Many persons were sleeping in the house—about 30, it is said; and there had been a children's party in it the evening before, when nearly a hundred were present; but almost all these were gone before midnight; after which both Mr. Paley and the butler had gone round the house and seen that all appeared safe.

———

January 5th.

Fanny was told to-day by the Loraines that the fire was *not* caused by a flue, but by a lighted candle left by some careless servant in a small closet. This also is certainly, a very possible cause of fire.

Read some of Green's History, also part of

1885. Alfred Newton on Birds, in Encyclopædia Britannica
—new edition.

_____

January 6th.

A beautiful clear sunny day with white frost. A
good walk with Laura and Fanny—the three dogs
very diverting. Lady Hoste came to luncheon.

_____

January 7th.

Another beautiful bright frosty day. Had a
pleasant walk with Fanny and Laura. Read Cicero
De Senectute—also Horace.

The cause of the fire at Ampton still appears
doubtful. There is a very copious and careful
narrative of the whole, in the *Bury Post.* It seems
that the children, or some of them at least, had a
narrow escape, for they were sleeping in a separate
part of the house which the fire would in a very
short time have cut off completely from the acces-
sible portion ; so that it is said if the fire had gone
on undiscovered a little while longer, they would
hardly have had a chance of escape. The *Post* also
says that the construction and arrangements of the
house, exposed it to much danger of fire—that there
was an excessive proportion of wood in the building,
and of various combustible materials within it. I
knew nothing of the house beyond a part of the
ground floor.

_____

January 9th.

Freda Loraine came to luncheon. Read Cicero
De Senectute. Wrote a little of my Reminiscences.

Arthur Lyell and his wife (Florence Chambers) arrived, also Clement.

Read Cicero De Senectute.

———

Fanny and Laura both very unwell. William Napier and his daughters Emily and Hester arrived, also Constance Wilson. Harry went to Bury to take the oaths as a Justice of Peace.

Read Cicero De Senectute.

— — ——

General and Mrs. Ives and their daughter arrived —also Cecil (the younger), young Parker, young Hoare.

Read Cicero De Senectute.

We have had a long talk with Mrs. Paley (Clara), and have had from her a very copious account of the burning of Ampton; but I still have not a very clear idea of the whole progress of the event. How the fire originated seems to be still quite uncertain, and I suppose will always remain so. It is clear that some of the children (the Paley's own son, and Lord Rayleigh's) who were sleeping in the top-most part of the house, had a narrow escape with their lives. Mr. Paley seems to have shown on this occasion a great deal of clearness of head and presence of mind, as well as courage and energy.

The newspapers are full of the rejoicings in

1885. honour of the "coming of age" of Prince Albert Victor. It seems to me but a little while since his *mother* entered London on her way to marriage."

Crabbe says,* old men are apt to remark that the days pass slowly with them, and yet the years, on looking back, seem to have gone rapidly. I find this true in some degree, as to the years, but not as to the days. I am thankful to say that, as yet, I am far from finding the days heavy or tedious, that is, when I am well in health and am here at home.

January 15th.

All our guests went to the Bury Ball.

Read Cicero De Senectute. An admirable work!

January 16th.

Dismal, dark and cold weather.

Read Proverbs, chapter iii. and iv. and v., with Dr. Plumptre's Commentary.

The Arthur Lyells and Clement went away. The Fancy Dress Ball at Hardwicke—our guests all went to it.

January 17th.

All our guests went away except the Napiers.

Read Proverbs, chapter vi. and vii., with the Commentary.

Fanny better—Laura still confined to her room.

---

* "Tales of the Hall," book 10, conclusion of "The Old Bachelor's Story."

The same detestable weather.

Read Proverbs, chapter x., with Commentary.
Read also Gibbon.

Lady Hoste and Mr. Greene, Dorothy Hoste,
Miss Drummond Wolff and Mr. Bentinck dined
with us.

———

Dear William Napier left us — his daughters
remaining. A visit from Mr. Harry Jones. Lady
Rayleigh, Mr. and Mrs. John Paley and their little
boy came to stay with us.

Read Proverbs, chapters xi. and xii. Read
Gibbon.

———

Very cold.

Read *Edinburgh Review* on Mallet du Pau, and on
Croker.

The Paleys and Lady Rayleigh went to examine
the ruins of Ampton.

Again a terrible battle in the Soudan, between
our force under General Stuart and the Arabs; we
have been victorious, but with most lamentable loss
of life: 9 officers killed (one of them Colonel
Burnaby) and 9 wounded; soldiers: 65 slain, 95
wounded. It is little consolation that we have
killed a great number of the poor Arabs, who seem
(as in the engagements last year) to have fought

1885. with most desperate and reckless valour.   I suppose they must be inflamed by fanaticism.

Colonel Burnaby, I should think, is a great loss; he seems to have been a thoroughly chivalrous character. All our troops who were engaged seem to have behaved admirably. But our whole force out there is so small that, unless the enemy are thoroughly disheartened, we cannot look without great anxiety to the prospect of another battle.

Talking of the fire at Ampton, John Paley told me of a curious fact he had observed :—that papers (old papers) resisted the action of the fire better than parchment; that in several instances, where papers and parchments had long been tied up together in the same bundle, it was found, after the fire, that the parchment was shrivelled and distorted so as to be utterly illegible, while the paper was hardly discoloured.

The origin of the fire is still quite uncertain.

John Paley told me that during the whole time of the fire, he seemed to himself to be in a state of extraordinary excitement,—almost of exaltation: he felt neither fear nor cold.

January 23rd.

A very fine day with sharp frost.   Had a pleasant walk with the Napier girls.

Read *The Quarterly* on Carlyle and on Johnson.

Lady Hoste, Dorothy and Miss Drummond Wolff came to luncheon and to see our pictures.

Walked in the garden with Fanny. Lady Rayleigh went away. Wrote to Louis Mallet. Read *The Quarterly* on the Highlanders and their Landlords.

Weather not quite as fine as yesterday.

January 26th.

A cold thaw and fog.

Read Proverbs, chap. xiii, also Gibbon ; also part of Miss Paget's Countess of Albany, and a little of Vita di Alfieri.

The shocking news came yesterday of the horrible crimes in London—the dynamite explosions at the Houses of Parliament and at the Tower. There can be no doubt that they are the work of Irish-Americans. Unfortunately the infernal villains have been more successful in doing mischief this time than in former attempts : two unfortunate policeman have been dangerously wounded, and some other people much hurt, and the interior of the House of Commons completely "*wrecked*," besides some mischief (but less) done at the Tower. Still the effect of these devilish attempts appears small in comparison with the design ; there seems something almost paltry in making such great efforts to damage a building when there were none in it but some poor innocent sight-seers. One must trust in God that they may never be more successful.

Read Gibbon, and went on with the Countess
of Albany.

News of another battle and another victory, but
not won without heavy loss—not indeed so heavy
numerically, as in the last battle, but what seems
very serious is, that General Stuart has received a
severe and (it is feared) disabling wound. The
advantage gained, however, appears much more
important than before, as our forces have reached
the Nile, and opened a communication with
Gordon's steamers, so as to gain a command of that
part of the river.

The object of the laborious and dangerous march
across the Desert was (I suppose) to cut off the
great bend which the Nile makes.

———————

February 2nd.

A fine day.
Walked with Fanny and Hester Napier.
Read Gibbon.
Wrote a little more of my Reminiscences.

———————

February 3rd.

Walked in the garden with Fanny and Laura.
Mrs. Wood arrived.

———————

February 4th.

A fine day.   Walked in garden with Fanny, Mrs.
Wood, Laura and others.

My 76th birthday. God Almighty be thanked for
all His mercies. I can only repeat what I said

on the corresponding occasion last year : acknow- 1885.
ledging with deep and humble gratitude the great
goodness of Almighty God, in allowing me to arrive
at such an age in good health and in the enjoyment of so many blessings.

Spent half-an-hour in my museum, first time
since November 9th.

———

February 5th.

Fine, with rough and cold wind Walked with
Mrs. Wood and Laura.

Read Max Müller on Bunsen.

———

February 6th.

Began to read Dean Church on Bacon.

Walked with Fanny.

Read part of " Milly and Lucy."

Emily and Hester Napier, George and Arthur
went away.

Most deplorable and disastrous news, yesterday,
by telegraph :—Khartoum betrayed by treachery
into the hands of the enemy; Gordon's fate uncertain, but there seems every probability that he is
either slain or a prisoner. Sir C. Wilson's force,
having attempted to approach Khartoum, repulsed
by the heavy fire of the enemy ; the Mahdi's force
there seems to be great. Much fear that Wolseley's
army will not be strong enough in numbers for the
work it has to do.—It is an awful crisis.

LETTER

Barton Hall,
February 6th, 1885.

My dear Leonora,

1885.　　　Many thanks for your kind letter and good wishes on my birthday. It is indeed a great cause of thankfulness that at this advanced age I find myself in such good health, and able to enjoy so many sources of happiness. I am very glad indeed to hear that Annie has recovered her health, and I hope she will have no more illness to interrupt her pursuits or to make you uneasy. I hope too that Dora is well. This excessively uncertain and changeable winter weather makes a great deal of care necessary to preserve health ; it is so difficult to avoid chills.

I am very glad you have read the Memoirs of Sir George Napier, and I am sure you would be interested by it, it is such a thoroughly true and natural picture of his own beautiful character— it is completely *himself*, exactly as I remember him and loved him ; and though it shows in strong colours the horror of war and the miseries inseparable from it, it also shows that a really grand and noble character like Sir George's can pass through such severe moral trials without being infected.

I read the life of Sidney Smith (by Mr. Stuart Reid), last November, and also the Life of the same by his daughter, Lady Holland—both are entertaining, but I like Lady Holland's the best.

I have now begun a little book on Bacon, by the 1885. Dean of St. Paul's (Church), in the series edited by John Morley.

Believe me ever your loving brother,

CHARLES J. F. BUNBURY.

JOURNAL.

February 7th.

A very fine day.

Had a pleasant walk with Laura, Fanny accompanying in her pony-carriage.

LETTER.

Barton, Bury St. Edmund's,
February 7th, 1885.

My dear Katharine,

Many thanks for your good wishes on my birthday, and for sending me that fine book of comical illustrations to the Jackdaw of Rheims.

Feb. (8th). I had written no further than this yesterday, and this morning I receive your interesting letter. The news from Khartoum is indeed deplorable, and not only in the present evil, but in what it leaves one to apprehend; for though Gordon individually may be the greatest loss, it is still worse to think of the destruction of the whole force under Wolseley as possible if not probable. It is dreadful to think of the grief and anxiety that so many must be suffering in this suspense. If

1885. it be true that Gladstone was against sending
Gordon on this wild-goose chase, he is much to
be pitied, for certainly everybody holds, and will
hold him responsible for the whole scheme. His
name will be associated with the expedition as
Lord Auckland's is with the first Affghan war; for
one looks on Gladstone as not only Premier, but
Dictator. I cannot go along with you in general
admiration of Bright, but I do think that he was
very wise in keeping out of this business altogether.

Why should we send expeditions to slaughter poor
ignorant Arabs or Egyptians, who would never have
meddled with us ? Then again, the dynamite
atrocities make one excessively uncomfortable, for
one dreads not only what may happen to one's
own friends, but also injuries to invaluable public
treasures. It is horrible to think that the British
Museum or the National Gallery might be destroyed
by those abominable villains—much worse savages
than those of the Soudan. What with war in
Africa and dynamite at home, Socialists in the
Ministry, and *no* price for farm produce, it is difficult
to be happy and comfortable in such days. However, as old Bewick's vignette says — "Good times
and bad times, and all times, pass over."

Fanny has given me, for a birthday present, a
volume of selections by Mrs. Kingsley,* from her
husband's writings, and I find great comfort in
them.

We are anxiously waiting from day to day, for the

* Daily Thoughts.

great event in our family which it seems cannot be far distant; but Laura still looks very well, and seems in excellent spirits. Fanny and I, I am thankful to say, are also very well.

Ever your loving brother,

CHALES J. F. BUNBUY.

JOURNAL.

February 9th.

Very fine day, very cold wind.

Walked with Fanny, she in pony-carriage.

Captain and Mrs. Lambart arrived, and Arthur returned from Sawbridgeworth.

Read some more of Dean Church on Bacon.

February 10th.

A very fine day.

A pleasant walk with Helen Lambart.

Mr. and Mrs. Robert Marsham, Mr. Kenyon, Reginald Talbot and John Hervey arrived.

The John Paleys dined with us.

February 11th.

A pleasant walk with Laura and Mrs. Wood.

Read Job, chapters xxxviii. and xxxix.

Major and Mrs. Harris, Miss Meade, Mr. Bevan and his daughter Mabel dined with us.

There seems to be now no doubt that Gordon was murdered when Khartoum was seized and given up to the enemy by the treacherous Pashas. And

1885. better that it should be so—that he should have died at once rather than have fallen into the hands of the enemy. He was a grand man—certainly a hero—a very extraordinary and memorable character; yet I cannot help wishing that he had died before he undertook—on behalf of our Government —the expedition to Khartoum. Or rather, indeed, it would be more reasonable and just to wish that our Government had never urged or persuaded him to undertake that wild expedition. However it came to pass, or whoever is answerable for the whole deplorable series of blunders, it is certain that we are (speaking in familiar language) in a dreadful *mess*, with no hope or chance of extricating ourselves unless by further cruel sacrifices of valuable lives.

LETTER.

Barton, Bury St. Edmund's,
February 11th, 1885.

My dear Joanna,

Very many thanks for your kind and agreeable letter, dated on the 4th, and for all your good wishes for which I am as grateful as if they had arrived on my birthday. I am thankful to say that Fanny and I are now both of us quite well, and able to enjoy the very fine weather which we have had for some little time past. The spring is so far a fine and forward one, and many pretty flowers are appearing in the garden :— Yellow Aconite (*Eranthus* properly) in profusion ; Snowdrops of two species in abundance, Violets, Yellow 1885. Crocuses, and the beautiful little blue Scilla bifolia. In the hot-house we have some very beautiful Orchids in blossom. The only place where I ever saw Snowdrops really wild, was on Monte Albano, the ancient Mons Latialis. I am interested by your account of your travelled Italian friend ; but I must own that Siberia and Lapland are almost the last countries in which I should wish to travel.

I lately read a Memoir of the Countess of Albany by "Vernon Lee," published in the " Eminent Women" series. It is entertaining, though I do not much like the style ; of course it sent me back to read Alfieri's Life over again, which is always entertaining. Fanny says that "Vernon Lee" is a Miss Paget, and that you know something about her.

I have lately read, for the fourth or fifth time, Cicero's treatise on Old Age (De Senectute) : it is beautiful. I am now reading a volume on Bacon by the Dean of St. Pauls.

All our thoughts are now very much occupied by the deplorable news from the Soudan ; I hardly think that England has been in such a distressing and alarming position since the first Afghan war in 1842 (or at any rate since the Indian Mutiny). The death of Gordon is to be lamented, but it is worse to anticipate the multitude of deaths that must follow—the multitude of valuable lives that will be sacrificed in vain. Even since I began this letter there comes the news of another murderous battle, with lamentable loss of valuable men. General

1885. Earle, Sir William Codrington's son-in-law, seem to be particularly regretted. In all the recent battles, the loss of officers in proportion to soldiers has been excessive, much beyond what was usual in former wars; the Arabs seem to be not only brave fighters but skilful marksmen, and they have such an enormous superiority of numbers, that they can afford to *expend* a great many more men than we can. Our soldiers fight like heroes, but it does seem deplorable that they should be sent to be slaughtered by, or to slaughter, other brave men with whom they had nothing to do, and who would never have come to attack *us*. I wish the members of the Ministry could be sent to fight in the Soudan!

We are anxiously expecting dear Laura's *trouble*, and most earnestly hoping that she may get well over it. As yet she looks very well and seems in excellent spirits. I am very fond of her.

Ever your loving brother,

CHARLES J. F. BUNBURY.

Pray give my love to Susan, and many thanks for her good wishes. I will write to her bye-and-bye.

JOURNAL.

February 12th.

A beautiful day; very mild.

A pleasant walk with Laura, Mrs. Wood, and Fanny.

More news from Khartoum—death of General Earle.

The Nether Hall party.—Mrs. Wilson and Ida 1885. and others dined with us.

February 13th.

A cold, dull, disagreeable day. The Marshams went away early. Reginald Talbot, Mr. Kenyon, and John Herbert, after breakfast. Finished letter to Joanna. An attack of lumbago and a bad night.

February 16th.

Servants' dance.

February 17th.

Laura's baby (girl) born.

LETTER.

My dear Edward,

The expected addition to the Bunburys has arrrived.—Dear Laura brought forth a daughter yesterday evening (between seven and eight I believe), and I am happy to say that both mother and child are reported to be " doing well." This is not merely the conventional phrase, but I just now saw Dr. Macnab after his visit to the sick room, and he says the baby is a fine, healthy and strong one. I shall be very glad when she can come among us again, for I am very fond of her.·

I have been suffering since last Friday from a sharp rheumatic attack (beginning with lumbago, afterwards tending to sciatica), which has given me a good deal of pain ; but I am nearly well now, only stiff and weak in the joints, and altogether inefficient. For my comfort, I hear that almost everybody about us has been suffering from rheumatic attacks in one· form or another.

I hope you have fared better in London and Brighton. Fanny, I am happy to say, is fairly well.

Ever your affectionate brother,

CHARLES J. F. BUNBURY.

JOURNAL.

Very good reports of dear Laura and her babe.
A visit from Lady Hoste. Fanny, in evening, read a canto of " Marmion," (" The Camp,") to Mrs. Wood and me—read it excellently.

Nursery report very good. I was better, but still very stiff, though without pain. Fanny read a canto of " Marmion," (" The Court ") to Mrs. Wood and me.

The bulletins of dear Laura's condition and progress are all that we could wish ; most comfortable and gratifying ; equally so of the baby.

Received very agreeable letters from Joanna and Edward. A brilliant day, but hard frost ; thermometer 21 degrees.

Convalescent.
Good report of dear Laura and the baby. Fanny in evening, read conclusion of " Marmion," to Mrs. Wood and me.

A beautiful day, really mild—I out—first time

1885. since 12th, enjoyed the air and the beautiful flowers.

— — — —

Much rain.

Lady Hoste and Dorothy came to luncheon. We arranged new books in the library for an hour.

Read (second time) Dean Church on Influence of Christ's Character in the Latin races.

— — — —

Mrs. Wood went away. We arranged books in the library for some time.

A visit from Mrs. Wilson—very pleasant.

— — — —

Nursery report satisfactory. We made good progress with the shelf catalogue of library.

— — — —

Morning mild. Walked in garden with Fanny. Afternoon—rain.

I, decidedly convalescent. Fanny went in the carriage to Ampton and Livermere.

— — — —

Very good report of Laura and baby.

We made good progress with the *shelf catalogue* of the library. Fanny read to me in evening some of Elaine.

I have been out of health ever since the 13th

February, suffering at first from painful rheumatism or sciatica, and since from general debility, with some derangement of digestion — altogether in rather a low, dejected, and useless state. The weather has been unfavourable to me.

March 4th.

Heavy rain in night (nearly half-an-inch). Day mild and fine. Walked in garden with Fanny.

Mrs. Wilson spent the afternoon with us—very pleasant.

March 5th.

Dear Kate Hoare and her husband; Barnardiston and Lady Florence, with a son and daughter arrived.

We made good progress with the shelf catalogue of the library—finishing book-case 6.

March 6th.

Dear Katharine arrived. The Loraines, Mrs. Horton, Mrs. Wilson and Ida, *cum plur. aliis*, dined with us.

Read a little of Gibbon.

March 7th.

Walked for some time in front of house, the day being fine and not annoyingly cold.

The Barnardistons went away. Lady Hoste and Mr. Green dined with us.

1885.   Wrote a botanical note from H. Johnston's paper on Kilimanjaro.

_____

March 8th.

Read papers with Fanny and Katharine—and afternoon walked with them in garden for nearly an hour, the day being fine.   Arthur's illness.

_____

March 9th.

Arthur suffering from sore throat—uncertainty.

We went on for some time with the shelf catalogue of the library.   Read Psalms 19, 20, 21, 22, with the Speaker's Commentary.

_____

March 10th.

Charles and Kate Hoare went away.   Katharine remained with us.   Arthur's malady turned out to be *scarlet fever*, though mild.   Great annoyance and anxiety—Arthur removed to gardener's cottage.

=======

LETTER.

Barton,
March 10th, '85.

My dear Edward,

We are in a dreadful mess—our old enemy, the *scarlet fever*, in the midst of us.   Arthur went up to London to attend the Levee; spent some days in town, and returned with the scarlet fever!—though it was not at first recognized as such : and for two or three days past we have been trying to believe that it was a mere sore throat : but this morning Macnab has pronounced it to be really our scarlet enemy. I wish it may be possible to insulate him as thoroughly as was done with young Newdigate; but at any rate it is a dreadful plague: especially with Laura and her baby in the house. 1885.

(March 11th.)—I had not time yesterday to finish my account of *our enemy* and of the precautions which have been taken (arranged by Fanny and Macnab) to prevent it spreading further. These will, I should hope, be effectual. Arthur has been removed and settled in the *gardener's cottage*, where there is good accomodation for him and his nurse and his servant; Allan being taken into *this* house. I hear that Arthur's attack is a very slight one, and that he is in very good spirits. He has not been anywhere near Laura since her confinement (being in quite a different part of the house), so we may hope there is no risk of infection for her or the baby. Still the annoyance is great: for besides the trouble Fanny has had in making the arrangements, we shall be in quarantine,—cut off from our neighbours and friends for I don't know how long: besides having to destroy or purify all the furniture of Arthur's rooms. Laura and the baby are both, I hear, going on very well.

The rheumatic or sciatic ailment from which I was suffering when I last wrote, did not last long, but it left (as I am told is very often the case), a strange degree of languor and debility; this also is gradually passing away, and I think I could very soon be quite well if the bitter east winds would allow me to take exercise.

1885.   I owe you many thanks for your agreeable letter of the 20th.   I entirely agree with you that the one bright spot in the present aspect of affairs is the feeling shown by the colonies.

With much love from Fanny.

Believe me ever,

Your affectionate brother,

CHARLES J. F. BUNBURY.

JOURNAL.

March 12th.

Signed lease of my house, No. 24, St. James' Square.

Walked a little in the garden with Katharine.

Began to examine Barton Estate Accounts of last quarter.

Dear Katharine went away—I very sorry.

March 13th.

Walked some time in garden with Fanny.

Received from Lord Stradbroke Harry's appointment to be Deputy Lieutenant—wrote to thank Lord Stradbroke.

March 15th.

A beautiful day—lounged in the garden with Fanny ; we read prayers together.

Saw the *baby*.

March 16th.

Was allowed to see dear Laura again—first time since 16th February; very glad to see her again.

Went on with Barton Estate Accounts.

March 18th.

Harry and Laura, with the baby and nurse, went away to stay with Mrs. Wilson at Langham.

We made good progress with the Library catalogue.

March, 19th.

Library shelf-catalogue, finished bookcase 14.

We read a little of Gordon.

March 21st.

We went on with the Library shelf-catalogue, finishing bookcase 15.

Lady Hoste came to luncheon.

March 23rd.

We got on capitally with the shelf-catalogue of the books, finishing the great Library and 3 shelves of the little Library.

Went on with my Reminiscences.

March 26th.

A long discussion with Fanny, Scott and Betts on Mildenhall business.

1885.   Walked in the garden—day fine, but cold.
A pleasant visit from Mrs. Wilson and Laura.

March 28th.

A very fine day—we spent an hour in the garden.
The Library catalogue continued.

March 30th.

Dear Minnie arrived to my great joy.
A beautiful bright day.
Scott's report of Mildenhall Rent Audit—satisfactory on the whole.

March 31st.

A beautiful and mild day.
Had a pleasant walk with Minnie.
In the evening read aloud some scenes of "Two Gentlemen of Verona," to Fanny and Minnie.

April 1st.

A pleasant little walk with Minnie.
We went on with the shelf-catalogue of the library.
Read *Nineteenth Century* on Ghost Stories.

April 4th.

We went on with shelf-catalogue of library (rare books).
Read aloud to Fanny and Minnie some more of "Two Gentlemen of Verona."

LETTER.

Barton,
April 4th, 1885.

My dear Susan,

I began a letter to you nearly two months 1885.
ago, to thank you for your kind wishes on my birthday; but my good intentions were before long
checked by a cold and rheumatism, which for a good
while rendered me very flabby and useless. I am
now, however, (I am thankful to say), quite as well
as I can reasonably expect to be at my age and
at this season; and I enter into these particulars
only because you have in some of your letters to
Fanny expressed some kind anxiety about me.

You have heard from Fanny all about our scarlet
fever troubles, and other disagreeables which cannot
so well be named, but which have kept us in a
perpetual state of worry: — coming upon us too,
just when our rents are reduced. Altogether, I
think this winter has been about the most uncomfortable that I remember for a long time. Yet it is
a comfort that dear Laura has kept so well, and I
hope that after she and her husband and baby are
once settled in the Cottage, we shall derive much
comfort from having them so near at hand.

Though I have been confined to the house much
of this winter, I have not read a great deal; indeed
I can give but very shabby accounts of my reading.

I am afraid I must say, like Gray, "the days and
the nights pass, and I am never the nearer to
anything but that to which we are all tending."

1885.   You all seem to be delightfully busy, and it is a
pleasure to hear from your letters to Fanny of all
your occupations.   She, too, is making capital
progress with the Memoir of your father, from
which she reads to me many interesting bits.   I
am very glad you are reading the Memoirs of Sir
James Mackintosh, and that you like it,—it is one
of my favourite books, and I have read it more
than once or twice.

I am reading (slowly) the History of the Early
Reign of Henry the Eighth, by Professor Brewer ;
there are interesting portions in it here and there,
but on the whole it is very heavy.   That period,
however, (to the death of Wolsey), is one of which
one gets but an imperfect notion from ordinary
histories.

I see announced the Autobiography of Sir Henry
Taylor (Author of Philip von Artevelde) which I
should think might be entertaining.   But on the
whole, I am rather getting into Macaulay's way
— of liking to read old favourites again and
again.

I do not like to enter on the subject of politics : I
see nothing cheerful in it ; I see that "the world is
out of joint," but (more lucky than Hamlet) I do
not feel or fancy that I am born to set it right.
I am wrong, though, in saying that I see nothing
cheerful in politics.   The good feeling and cordial
friendly spirit which so many of the Colonies are
showing, are exceedingly cheering and satisfactory,
and I trust that the good spirit will be reciprocal
and permanent.   I agree with you that we ought

not altogether to despond : we have been in worse 1885.
scrapes before—in the first Afghan war for example.
Yet it does make me unhappy and angry that so
many noble and precious lives should be sacrificed
for no good object or purpose at all that I can see.
And I fear that a great many more will be sacrificed
before the summer is over.   Nor is it any great
comfort to know that we have slaughtered a great
many of those poor brave barbarians, who are
fighting for their independence.

I hope you have pleasanter weather at Florence
than we have here ; it is rather worse, I think, even
than our usual springs.

Much love to all your party.

Ever your loving brother,

CHARLES J. F. BUNBURY.

JOURNAL.

April 6th.

Harry and Laura came to establish themselves
(at last) in the Cottage ; they drank tea with us.

I began to read Wordsworth's Prelude.   Read
Mr. Pattison on Erasmus in Encyclopædia Britannica.

April 7th.

Mr. Harry Jones came at 10.45 and administered
the Holy Communion to Fanny and me and two of
the maids, in the dining room.

1885.   April 8th.

Wrote to Edward.

Went on with Wordsworth's Prelude.   We went on with the shelf-catalogue of the Library.

Read aloud in evening some scenes of "Twelfth Night."

April 9th.

My Barton Rent Audit :—the tenants very civil, though in low spirits.

April 10th.

Very wet and cold weather.
Clement arrived.
Resumed my Notes on Mildenhall Plants.

April 11th.

A long visit from Lady Hoste.
Arthur allowed to walk in the garden.

April 12th.

Weather mild but very damp.
Enjoyed the garden and lawn, after confinement to the house since the 2nd.
Read prayers with Fanny.
Read Psalms 42, 43, 44, with the Commentary.

April 13th.

Arthur released from his long quarantine in the Garden Cottage, ever since March 10th.   Set off for

Felixstowe. We had the telegram of his safe 1885. arrival.

I walked in the garden with Fanny and Minnie.

April 14th.

Had a pleasant walk with Minnie.

Received a charming letter from Susan Horner.

Went on with the Prelude, and re-read part of Dorothy Wordsworth's "Tour in Scotland."

April 15th.

We went down to the Cottage and saw dear Laura and Harry very busy and happy arranging their pretty home—also the dear baby. God grant them a long term of happiness there.

April 16th.

We went on with the shelf-catalogue of the Library.

I went on with the Prelude and Dorothy Words-worth's "Tour:" also wrote a little of my Reminiscences and of my Notes on Mildenhall plants.

April 17th.

Walked with Fanny; we inspected the new plantation near the East Lodge. We went on with the Library catalogue.

April 18th.

A beautiful, brilliant day.

1885.   Took a pleasant drive with Fanny in the open
carriage—to Bury one way and returned another.

Laura and Harry drank tea with us.

A visit from Lady Bristol, Lady Mary and Lord
John Hervey—pleasant.

Began to read Autobiography of Sir Henry
Taylor.

Another splendid day.

Dear Mrs. Wilson and Agnes, Harry and Laura
came to luncheon.

Lord John Hervey, Lady Hoste, Mr. Greene,
Mr. and Mrs. Claughton, Mr. Abraham and his
daughter Louisa dined with us.

We (Minnie included) drove to Ickworth and saw
Lord and Lady Bristol, Lady Mary, Lord John
and Lord Francis—all very pleasant and friendly.

Beauty of conservatory and garden at Ickworth.

Spent a little time in my Museum, where I had
hardly been since November.

We went on with the Library catalogue.

Read some more of the Prelude, and of Sir
Henry Taylor.

Read aloud in evening, Act 1 of " The Taming 1885. of the Shrew."

Walked with Fanny down to the Cottage, and saw dear Laura and the baby.

We went on with the library catalogue ; finished the shelves in my study. Read aloud in evening Act 2 of " The Taming of the Shrew."

Arthur came back to us from Felixstowe, where he had been in quarantine after scarlet fever.

Finished Vol. 1 of Autobiography of Sir Henry Taylor.

A very hot sun with cold wind. Read prayers with Fanny. Began Vol. 2 of Autobiography of Sir Henry Taylor.

Mrs. Mills arrived. I had a very bad night.

Arthur set off for Dublin to rejoin his regiment.

Had a pleasant drive with Fanny—saw the ruin of Ampton.

Laura, her sister Caroline, and Harry dined with us.

Dear Minnie returned from London, and her brother with her. We were very glad to see them.

1885.   May 2nd.

Edward arrived. I had a pleasant short drive
with Fanny.

_____

May 3rd.

The Christening of Harry and Laura's dear little
baby—I, Godfather, Fanny acting Godmother,—
Mrs. Wood, Minnie, John Herbert and others
present.   _____

May 4th.

Poor Mrs. Mills suffering from neuralgia. Mr.
and Mrs. Bland, Mrs. Wood, Mr. William Wood,
Harry and Laura dined with us. Finished reading
Autobiography of Sir Henry Taylor.

_____

May 5th.

Mr. and Mrs. Saumarez came to luncheon, and
we took them to see the garden. Began to read
George Eliot's Life.

_____

May 6th.

Edward went away. Mrs. Wood, Mr. William
Wood, Laura and Harry dined with us. In the
shelf catalogue of the library, we finished the billiard
room.

_____

May 7th.

Mrs. Mills better. Captain Anstruther dined
with us.

_____

May 8th.

John Herbert went away. Fanny ill with cold
and cough

Lord John Hervey dined with Mrs. Mills, Minnie
and me

May 9th. 1885.

Fanny better. Read much of George Eliot's Life, and a little of the Geology of the Montreal Meeting Report.

———

May 12th.

Went out (first time since 4th) for half-an-hour in front of house with Minnie—the day beautifully bright, but wind cold.

———

May 13th.

Our dinner party.—Mr. and Mrs. Claughton, Mr. Abraham and two daughters, Lord John Hervey, Colonel Browning, Arthur Wilson, Francis Anstruther, Mr. Bence, Harry and Laura, Mrs. Mills, Minnie and myself.

———

May 18th.

Archdeacon and Mrs. Chapman with two daughters, and Mr. and Mrs. Barber arrived to an early dinner, and they and Minnie went to Fanny's festival for the drovers at Bury. Fanny could not go, but it was very successful.

———

May 19th.

Mr. and Mrs. and Miss James and Lord John Hervey, dined with us.

———

May 20th.

The Chapmans and Barbers went away.
We returned to the shelf catalogue of the library ; went on reading George Eliot's Life.

———

May 21st.

Louis and Fanny Mallet arrived.

LETTER.

Barton, Bury St. Edmund's,
May 21st, 1885.

My dear Katharine,

1885.   Very many thanks for your very pleasant letter received this morning. I am happy and thankful to say that Fanny seems to have rapidly and almost completely recovered from her severe cold and cough, which for some time made me very uneasy; to-day she has been out in the garden, for the first time for more than a fortnight, and has enjoyed it, and does not seem to be the worse, though the wind is still very cold. I thoroughly sympathize with you in your remarks on the season. How very disagreeable it is of the floating ice to come in such quantities into our seas and make us so cold! And yet, with all this cold, the trees and grass and the garden flowers are in full beauty, and if one only looks through the windows, it is often difficult to believe that the outward air is so icy. The lilacs are in the fullest beauty, and the fruit trees also, especially the apple blossom, which I have never seen more beautiful. The rhododendrons are not yet quite out, but in that department you know we are poor. In the hothouse, we have just now some exquisite Orchids in flower :—Odontoglossum vexillarium (?—I am not quite sure of the name), Calanthe veratrifolia, Dendrobium infundibulum, and (finest of all) Sobralia macrantha. Of Puya I know nothing, except from Bentham's and Hooker's "Genera Plantarum," and the par-

ticulars there given do not give one any very dis-
tinct idea of it.

I am interested by what you tell me concerning
Miss North ; I am very sorry she is ill, but the only
wonder I think, is that she has not been dead long
ago. What she has done—both in travelling and
painting, is perfectly marvellous.

I am reading George Eliot's Life, and am now
near the end of the second volume. The greater
part of the first volume I did not like at all, but it
improved very much (to my thinking), after she
became a novel writer ; and much of what she
writes, both about natural scenery and about works
of art, is very interesting. Still I think it is much
too long, and I get almost as weary of her state of
health as I was of Mrs. Carlyle's. Sir Henry
Taylor's Life I found very entertaining.

I have re-read (for the 3rd or 4th time) Sara
Coleridges Letters : one of my favourite books.
Also I have read Wordsworth's White Doe of
Rylstone, which I think charming.

I am tolerably well, free from any pain or positive
ailment, but still far from strong ; I doubt whether I
could walk a mile without stopping to rest : and
my appetite is poor, and I can hardly hope to get
quite well till the weather improves. I do not know
how I can get to London.

Ever your loving brother,

CHARLES J. F. BUNBURY.

JOURNAL.

1885.   Willie Bunbury and his wife and their dear little boy arrived.

———

Violent storms of rain, with bright intervals. A pleasant little walk with Minnie. Harry and Laura and her brother dined with us.

Wrote a little of my Reminiscences.

———

The Willie Bunburys, with their child, went away after a tree had been planted for little Gerald.

———

Walked in the garden.

A visit from the dear baby. Spent some time in my Museum.

Read Sir H. Thompson on *Diet.*

———

A most beautiful day. Had a delightful drive in the open carriage with Fanny and Louis and Fanny Mallet.

———

Another splendid day, very hot. Had a very pleasant drive with Fanny Mallet.

Louis and Fanny Mallet went away: we went with them as far as Bury.

Norah Aberdare and her daughters Lina and Lilly arrived.

Read Lord Acton on George Eliot.

May 30th.

The 41st anniversary of our happy marriage. Thanks be to God for all our blessings.

Had a delightful walk with dear Norah.

Lord John Hervey arrived. Harry and Laura dined with us.

June 1st.

Dear Norah Aberdare, with her two nice girls, and Mrs. Mills went away.

Walked with Fanny and Minnie to the farm : saw Mr. Scott and the young donkey.

June 2nd.

A most beautiful day. We (3) made a delightful expedition to the Shrub, where the *blue-bells* gloriously beautiful. We called on Mrs. King.

June 3rd.

We had a pleasant drive round by Pakenham and Norton, all the country in general beauty.

June 5th.

We had a delightful lounge in the arboretum in

1885. morning, and a pleasant drive in afternoon.   Harry,
Laura and Mary Wood dined with us.

_____

June 6th.

Dear Minnie left us in the morning.

Leonora and Annie Pertz arrived—we very glad
of their arrival.

_____

June 8th.

Dear Leonora and Annie went away after
luncheon.

A pleasant visit from John Paley.

_____

June 9th.

Morning wet, afternoon beautiful.

The startling news of defeat of Ministry and
expectation of their resignation.

A pleasant drive with Fanny.

_____

June 10th.

A very fine bright day.   We had a drive in the
pony-carriage, and strolled in garden.   Preparing
for move to London.

Reading "Colonel Enderby's Wife."

_____

June 12th.

A splendid day—excessively hot.   We went with
Scott through the Vicarage Grove, inspecting the
trees and shrubs.

_____

June 13th 1885.

Mrs. John Paley and her boy and Miss Paley came to tea, and we walked through the garden with them.

———— —

June 15th.

Up to London to 48, Eaton Place, by the 3.20 train from Bury: had a good journey: found Minnie established at 48.

———— —

June 16th.

A northerly wind, very cold.

Fanny took me to the Athenæum. I met Edward there.

————

June 18th.

Went out driving with Fanny : saw the beautiful Rhododendrons. Dear Cissy came and spent part of the afternoon with us.

Wrote some of my Reminiscences.

Charming Mrs. Storrs dined with us — also Clement.

————

June 19th.

Another drive in the Park with Fanny ; we called on Mrs. Wilson.

Charlotte and Octavia Legge came to luncheon.

————

June 20th.

We drove to Annie Pertz's studio ; saw the portrait of Sarah, which she is at work upon.

1885.   Our dinner party—Sarah and Albert, Susan and
Joanna Horner, Harry and Laura, very pleasant.

_____

June 21st.

Visits from Willoughby and Mrs. Burrell, and
Sir Alexander and Lady Wood.   The Palmer
Morewoods and John Herbert dined with us.

_____

June 22nd.

We went to the museum of Natural History—saw
statue of Darwin.   Also some of the stuffed birds.

William Napier and his daughter Emily came to
luncheon and Emily stayed with us.   Willie and
Mimi Bruce dined with us—very pleasantly.

_____

June 23rd.

Isabel Locke and Edward dined with us.

_____

June 24th.

A good report of my health from Dr. Harper.

We lounged for a little while in the Royal
Botanic Society Garden—very pleasant.   Visits
from Cissy and Emmy, Mrs. Walrond and Dora
and Maggie, Mrs. Martineau, Miss Adeane.

_____

June 25th.

Our dinner party — Mrs. Willoughby Burrell,
Edward, Harry and Laura, Admiral Spencer,
Captain and Mrs. Lambart.

_____

June 26th.

Our dinner party—Courtenay and Lady Muriel
Boyle, Clements and Mrs. Markham, Bernard
Mallett—all pleasant.

Dear Rose Kingsley came to luncheon with us— we very glad to see her—she delightful as ever. Visits also from Joanna (very pleasant) and some of the Wilson girls.

June 28th.

Susan Horner and Edward dined with us.

June 29th.

We drove to Kensington, visited dear Mimi Bruce, and saw her charming little children, Fox and Norah. I had visits from Rosamond Lyell, Susan MacMurdo, and Katharine—very pleasant.

June 30th.

We visited the French Gallery of Pictures. Sally and John Herbert dined with us.

July 1st.

We called on Mary Powys, and saw her; afterwards we went into the Natural History Museum, and went through the gallery of birds, seeing the splendid collection. Cissy and Emmy came to luncheon with us, and afterwards dear Sarah came to see me.

July 2nd.

A visit from Kate Hoare—she went with us to Annie Pertz's studio—we found Sarah there, and brought her back.

July 3rd.

We again visited Mimi Bruce, and afterwards
Katharine and Rosamond. Our dinner party—Mr.
and Mrs. Lacaita, the MacMurdos (3), Mr. King-
lake, Edward.

_____

July 4th.

Mrs. Young came to luncheon.   Sarah dined
with us.

Visit to Miss Sulivan at Fulham ; a beautiful
spot on the bank of the Thames, with an abundance
of beautiful trees, and profusion of interesting shrubs
and other plants. In particular, a variety of *bog*
plants, many rare ; numerous and fine plants
of one of the N. American Cypripedia (album ?
speciosum ?) now just passing out of flower :—
Gentiana asclepiadea, not yet in flower, but in
fine condition.

_____

July 5th.

To the Zoological Gardens—in great beauty.
We attended especially to the wading and swimming
and some of the Gallinaceous birds, which were
very numerous, lively and noisy ; the Pelicans,
Gulls, Flamingos and many others.   A rather small
Penguin was particularly tame and amusing,
waddling, or *toddling* in a very erect posture, after
its keeper and thus following him over an extensive
piece of grass.   The Boat-bill a very extraordinary
and uncouth bird.

_____

July 6th.

We drove to Rose Bank and saw Susan Mac-
Murdo, but Montagu was just going out.

Our dinner party—Willoughby and Mrs. Burrell, 1885.
William and Mimi Bruce, Mrs. and Miss Ives,
Mr. Elliot, Mr. Cavendish, our four selves at
home.

_____

July 9th.

Went with Fanny and Emily Napier, and saw the
" Old London Street " in the Improvements
Exhibition ; very clever imitation of portion of
London as it was in the old time, giving us a lively
idea of the old houses, streets, pavements, shops,
and general arrangements. It appears to be very
well done.

To the Natural History (British) Museum ; spent
some time very pleasantly in looking at the beau-
tifully-stuffed and arranged specimens of British
birds with their nests, eggs, and young ; arranged
mostly by Lord Walsingham. (I noticed them
in my Journal of last year).

_____

July 10th.

Splendid weather. We went with Emily Napier
to the Botanical Society's gardens in the Regent's
Park. These are very pretty and very pleasant :
—we had visited them on June 24th, when they were
in rather greater beauty, with many kinds of Iris
in blossom, and various trees clothed in their young
leaves. Now the Iris flowers are all past, and
the foliage of the trees more uniform ; but still the
general effect of the gardens is very pretty, and
many ornamental flowers to be seen.

We dined with Katharine, in Cornwall Gardens :
—a very pleasant party. Met Mr. and Mrs.

1885. Peabody, Mr. Leigh Smith (the famous Arctic voyager), William Nicholson and his daughter, Susan Horner, Edward.

Mrs. Peabody is a young American lady, very pretty, very intelligent, agreeable, and interesting; I met her two or three years ago (at the Leonard Lyells) and was much fascinated by her; since then she has married her cousin of the same name, and has just now re-visited England.

I was very glad to meet her again, and found her as charming as ever. Her husband, a very young man, not remarkable in appearance or manner, is said to be greatly devoted to schools and other good works.

_____

July 11th.

We visited Katharine, also Willie and Mim Bruce.

Our dinner party:—Lady Charlotte and Lady Octavia Legge, Mr. and Mrs. Heathcote, Mr. Walrond, Harry Bruce, John Herbert.

_____

July 12th.

Read prayers with Fanny.
Susan Horner and John Herbert dined with us.

_____

July 13th.

We three went to the Royal Academy, and spent some time pleasantly enough in looking at many of the pictures. Some of those I noticed were:—
Sinodun Hill, near Dorchester, by _Vicat Cole;_ —

A wild rock scene on the North Coast, by *Peter* 1885.
Graham ;— The Ruling Passion (a sick or dying
man, eagerly examining some specimens of natural
history), by *Millais ;*—A treatise on Parrots, by
Marks;—A reading from Homer, by *Alma Tadema.*

Sarah Seymour and Katharine Lyell dined with
us—very pleasant.

July 14th.

William Napier went with us to Miss Sulivan's
garden party at Fulham. A brilliant day, and a
very gay and pretty scene (see July 4th), the locality
particularly well suited for such a meeting :—a great
many persons whom we knew.

William Napier and Edward dined with us.

July 15th.

A beautiful day (though St. Swithin's).—To the
National Gallery, to see the two new pictures—the
Blenheim Raphael and the Vandyck. The Raphael
is in his earlier manner, and I am afraid I do
not altogether appreciate it, except the Bambino,
who is lovely.— *Vandyck's* Charles the First—the
King a fine and interesting portrait—the characteristic type of face which is so well known from
the numerous copies and repetitions—melancholy
and strikingly interesting. Various opinions as to
his horse.

July 16th.

Mrs. Storrs came to luncheon.

Our dinner party—Lady Winchelsea, Sarah Sey-

1885. mour, Willie and Mimi Bruce, Harry Bruce, Mr. Walrond, Mr. Sinclair, Mr. Marsham, Mrs. Wilson and Agnes, Edward, Clement, Minnie and Emily Napier.

_____

July 17th.

We visited Arthur Lyell and his very pretty wife —also Lady Head—also went in for a little while to Mrs. Douglas Galton's afternoon party.

_____

July 24th.

Our dinner party—Lady Alfred Hervey, Lady Hoste and Mr. Green, Dorothy Hoste, Sarah and Albert Seymour.

Mr. and Mrs. Henry Finch-Hatton, Mr. and Mrs. De Grey.

_____

July 26th.

We drove to Rose Bank, and saw a large and pleasant family party of the MacMurdos.

_____

July 27th.

The morning very hot, though the east wind continuing as for a long time past; towards evening same wind becoming very fresh, at last really cold.

We visited the National Portrait Gallery; very interesting. A large picture very lately given by the Emperor of Austria—representing the House of Commons in 17—, with Pitt speaking; the members all portraits.

Another large picture, showing the first House of Commons after the Reform Acts, in 1833 — all portraits, as in the other case.

To the Natural History (British) Museum. We went rather hastily through the gallery of minerals; this is really a magnificent collection, indeed I can well believe it to be the finest in existence, and not less admirable in arrangement and in exposition.

Afterwards we spent a good deal of time very pleasantly in studying the beautiful special set of British birds with their nests and eggs.

July 29th.

Car MacMurdo came to stay a day with us.
Sally, Clement, and young Cecil. dined with us.

July 30th.

A beautiful day. My old friend, John Carrick Moore, came to luncheon; I have a great liking and regard for him, yet this is the first time I have met him this year; indeed, I think he spends most of his time in the country. My acquaintance with him began in 1845, at the British Association Meeting at Cambridge; and the agreement of our tastes for natural science, drew us much together for many years. Some branches of zoology were favourites with him, while botany was my hobby; geology had attractions for both, though he was most devoted to it, and had pursued it with most success. He is a well informed and accomplished man, as well as a very pleasant man; I have always found him remarkably genial and pleasant, as well as a cordial friend.

We visited the Zoological Gardens, where the

1885. birds were in great glory, much enjoying the sunshine. I noticed particularly what I had never seen before— the Flamingos *swimming*. Though their webbed feet are very apparent, I had not before seen them entirely trusting to the water. I had supposed them to be *waders* more than swimmers.

The dear Seymour boys, Charlie and Bill, went with us, and with Minnie to the Zoological Gardens and were very pleasant and very amusing.

August 1st.

We visited Mrs. Mills, who is looking well.

August 2nd.

Captain Campbell, William and Mimi Bruce, and Captain Hervey, came to luncheon.

August 3rd.

The MacMurdos, Mrs. Clements Markham, Mrs. Frederick Campbell, Charlie Seymour and his little brother came to luncheon. Visits from Mr. Robert Marsham, Madame de Bunsen.

August 4th.

We took leave of dear Minnie—very sorry to part from her—she has been (as she always is) during all this stay in London—a most pleasant and kind and true *sister* to us.

We came down to Barton by the afternoon train by Cambridge. Leonora Pertz and her daughter Dora with us.

We arrived safe and well at home. Thank God.

Some heavy showers—very much welcomed by gardeners and farmers. The hard, parching drought which has lasted so strangely long, seems likely at last to break up.

Had the great pleasure of again seeing dear Laura and Harry, and being heartily welcomed by them ; they seemingly in high health and spirits ; their lovely baby too, looking fresh, fat and blooming.

I see announced in the newspapers, the death of my old friend Lord Houghton (Richard Monckton Milnes). He was very nearly of the same age as I —both born in the same year, 1809—but he by some months the younger ; on the other hand he was my senior in standing at Cambridge by (I think) two years. I do not exactly remember when I first knew him, but it must have been (I think) either in my second or third term, and through the medium of Arthur Hervey. From that time we continued to be intimate till he took his degree ; Edward and I being in some measure in the same *set* with Milnes. His acquaintance was much more extensive than mine or even Edward's ; but my especial friends—Stafford O'Brien, Augustus Fitz Roy, James Colville, were also, I think, friends of Milnes.

Milnes was a man of extraordinary activity and versatility of mind, with a great variety of pursuits, but hardly, I think, first rate in any. He spoke often at the Union, with great fluency and confidence, but hardly (I think) with brilliant effect.

1885. Confidence indeed was one of the qualities most
prominent and conspicuous in him; every one
knows the name which Sydney Smith gave him, of
" The Cool of the Evening." But with various
oddities and perhaps weaknesses, he was a truly
good-natured, and I believe, good-hearted man.

After leaving Cambridge, I saw Milnes only
occasionally, at uncertain and often long intervals;
and this was still more the case after we, each of
us married and settled. But whenever we have
met of late years, he has always greeted me in a
friendly and cordial way.

_____

August 6th.

We had a pleasant drive in the open carriage
with Leonora, and in the evening walked through the
garden and arboretum.

_____

August 7th.

A thunderstorm in morning; afterwards took a
pleasant drive in the open carriage, and strolled
through the gardens.

I had a bad night.

_____

August 8th.

Fanny read part of "She Stoops to Conquer,"
aloud in the evening—entertained us very much.

I had a better night.

_____

August 9th.

Laura and Harry, Cissy and Emily and Mr.
Burke came to tea.

Harry and Laura, Cissy and Emily dined with us—pleasant: Laura particularly pleasant.

A blowing day, though fine; we drove with Cissy and Leonora.

Read part of Vol. 1 of Adam Bede.

Had a very good night.

August 11th.

We took a drive with Cissy. Went on reading Adam Bede.

August 12th.

A pleasant drive alone with Fanny. Read the Book of Judges, chapters i. and ii.

August 14th.

A very pleasant drive with Fanny and dear Mimi Bruce. Edward arrived. Cissy and Emmy dined with us. The two Bruce children, Fox and Norah, very pretty and very amusing.

August 15th.

Splendid weather; the harvest going on gloriously. Indeed the weather has been fine, in the sense of not being wet, almost ever since we came home; but till the last few days it has mostly been rough and blustering, not agreeable. The last two days thoroughly fine, all that could be desired for the harvest; and the "happy autumn fields" are beautiful to look at. To-day we observed several of the fields towards Fornham already cleared.

We drove with Leonora and Edward to Ampton, and saw the ruins. Fanny had a visit from Lady Bristol.

The harvest is going on capitally; indeed all the better because there has been no oppressive heat to annoy the men.

Laura's baby is really a thorough beauty—one of the most lovely little creatures I have ever seen. It is a delight to me to see it.

And Laura continues to be as charming and love-able as I thought her last year, and Harry's and her marriage to be as great a blessing as I hoped it would be.

William and Mimi Bruce are with us now, as delightful as ever; and their two children, Fox and Norah, are lovely.

August 19th.

Mr. and Mrs. Livingstone arrived. Lady Hoste and Dorothy and Mr. Greene dined with us, as well as Harry and Laura and Mr. and Mrs. Byron.

August 20th.

Edward went away. Heavy showers. Read Judges xi. and xii.

A tolerable night.

August 21st.

The Livingstones went back to Mildenhall. We drove with Leonora to Bury and round by Hard-wick.

August 24th.

Harry and Laura dined with us. Read Judges xv. and xvi.

A good night.

––––––

August 25th.

A beautiful day, very warm. A pleasant drive with Fanny and Mimi, by Pakenham to Ickworth.

A visit from Sir Alexander Wood—very friendly, cheerful and pleasant.

A particularly good night.

––––––

August 26th.

Uncomfortable news from Edward—and doubtfully comfortable about Arthur. A long visit from Lady Gage.

Read Judges xviii.

Wrote to Edward. In evening agreeable recitations by William Bruce.

––––––

August 29th.

The Thornhills, with one daughter, dined with us. Spent a little time in my Museum.

––––––

August 30th.

We walked with Leonora in the garden and arboretum.

Read Haweis's sermon on Influences; read also the last two chapters of Book of Judges.

––––––

August 31st.

We took a drive with Leonora.

1885.   Read chapter 1 of the Book of Ruth.

Mrs. Walpole came to tea, and Lord John Hervey dined with us as well as Harry and Laura.

I have lately read a MS. Journal of the late Mr. Mallet, lent to us by his son Sir Louis Mallet. It contains much matter relating to English politics during the times of the ministry of Canning and Lord Goderich, and the early part of the Duke of Wellington's. Mr. Mallet appears to have been a very earnest, grave, thoughtful, sagacious man, with great knowledge and experience of political affairs : a close observer and rather severe judge of public men, with a strong leaning to republicanism, and accordingly, much inclination to judge English politicians severely. He scarcely seems to appreciate even Wellington or Peel :—Canning better, after his death. A politician who occupies a good deal of space in this journal is Herries, on whom, I should suppose, scarcely any one now bestows a thought. To be sure, he does not *here* appear in by any means a favourable light.

The steady, persevering continuance of the same weather through nearly the whole of this month (with the exception of a few days) has been very remarkable. On the 5th and on the 12th, and also on the 19th, 20th and 21st, there were heavy showers ; but otherwise the weather has almost constantly been dry, harsh and cold, with high winds almost constantly from north or east or some intermediate quarter. In fact, the prevailing weather during this August has been like that characteristic of March.

This pertinacious drought has been very favour- able to the harvest, which has gone on brilliantly, with hardly a day's interruption, and is now, in fact, nearly finished in this parish.

September 1st.

We went (Leonora with us) to Fornham, and saw Mr. and Mrs. Gilstrap (who were very courteous), and their fine collection of coniferous trees and shrubs. The old Lebanon Cedars very grand.

A very good night.

September 2nd.

We drove to Horringer and called on Sir Charles and Lady Ellice.

Read the remainder of the Book of Ruth.

I had not a good night.

September 3rd.

We drove with Mimi to the Livermere cottage, and saw Mrs. John Paley.

Mrs. Saumarez and old Miss Broke came to tea. Read 1st chapter and part of 2nd of 1 Samuel.

September 4th.

Lady Grey arrived. A good night.

September 5th.

Arthur arrived on crutches, poor fellow, and looking very ill, after his rheumatic fever. He had travelled from Dublin without stopping.

1885.   Mr. Abraham and his daughter and the John Paleys dined with us.

A very good night.

———

I went to Church (first time I have been there since May 3rd), with Fanny, Leonora, Dora, and the Bruces.   Afterwards walked in the garden.

———

Dear Willie and Mimi Bruce with their darling little children left us—much to our sorrow.

Read 1 Samuel, chapters iv. v. vi.

Resumed my Reminiscences.

Clement and young Cecil arrived.

———

Went on with my Reminiscences.

———

Read 1 Samuel, chapters vii. viii.

We drove (Lady Grey with us), to Hardwick, and saw Gery Cullum.   We went with Lady Grey through the garden.

I had a very good night.

———

We drove with Lady Grey to Langham, and saw Mr. and Mrs. Byron; in afternoon walked with Lady Grey to the cottage, and saw dear Laura and the baby.

Miss Wilmot arrived.

September 11th.   1885.

Lady Grey went away.

Went on with my Reminiscences.

———·———

September 14th.

Gery Cullum with two ladies, and two Mr. Crowes (brothers, a connoisseur and an artist) came to luncheon.

Mr. and Mrs. Byron, Ida Wilson and one of her sisters, Mr. Bevan and daughter dined with us.

——·———

September 15th.

A most beautiful day.

Dear Leonora and both her daughters left us (Leonora had been with us and Miss Wilmot in a drive in the morning).   Read 1 Samuel, xiii. xiv.

——— ———

September 16th.

Miss Wilmot went away.   We had a pleasant drive with Susan.   Spent a little time in my museum.

I had a very good night.

═══════

LETTER.

Barton,
September, 17th, 1885.

My dear Edward,

I was extremely glad to get your letter yesterday, and to hear that you have so far re-

1885. covered from your attack, and are so nearly convalescent. I hope and trust that you will, in a very few days, be able to get down to Brighton, and that the air there will soon completely set you up. We are now nearly alone : Susan Horner and Arthur MacMurdo with us, and no one else. Poor Arthur is very much crippled, and looking very ill. I do not know whether you are aware that, in the course of the last time he was at Dublin, he was very seriously ill—dangerously ill—with rheumatic fever, I believe—much more ill than we were aware of at the time ; he is now here on sick-leave, and it will be long before he will be able to join his regiment. He is looking excessively ill—very much reduced, and has lost several stone in weight : but he is under medical treatment and good care, and I trust will in time be quite right. Fanny and I are tolerably well.

The weather here is very variable—some days beautiful—others very wet. Nearly half-an-inch of rain (0.42) in the last 24 hours. I am very glad it is so fine with you.

With much love from Fanny, believe me ever,

Your affectionate brother,

CHARLES J. F. BUNBURY.

JOURNAL.

September 18th.

Montague MacMurdo arrived.

We had a drive with Susan. Received and began to read Life of Frank Buckland.

A visit from the Duke of Grafton.

Read 1 Samuel, chapter xvii.

We went on with Life of Frank Buckland.

Montague MacMurdo went away—we very sorry.

Read 1 Samuel, chapters xviii. xix.

A bad night.

A most beautiful day.

A pleasant drive as usual with Fanny and Susan Horner.

Had a really good night. Thank God.

Laura and her sister Ruth dined with us—very pleasant.

Wrote a little more of my Reminiscences.

Went on with Frank Buckland.

Finished reading Life of Frank Buckland.

Wrote to Joseph Hooker.

Drove with Fanny and Susan.

A very good night. Thank God.

Agnes and Ida Wilson arrived, and Mr. Hubert Hervey.

Mr. and Mrs. Holden dined with us.

———

Rain.

My Barton rent audit—deficiencies—but all civil.

Harry and Laura came to the luncheon party, and the Wilson girls and Arthur also joined in it, as well as Fanny and Susan.

A very good night.

———

Dear Agnes and Ida Wilson went away, to our sorrow.

Read General Walker's address on Geographical Section of British Association at Aberdeen.

Mr. and Mrs. Bland and the John Paleys dined with us.

———

Read prayers with Fanny and Arthur, and walked in garden with them and Susan.

Hubert Hervey went away.

———

Went on with my Reminiscences.

Read part of Edward Forbes' Natural History of the European Seas.

Nellie and Ida Wilson arrived; also Cecil the younger and Mr. Rycroft.

———

October 6th.

We went in the carriage to Hardwick with Susan —left her there till afternoon ; she went with Miss Robertson to Coldham Hall.

Began to read Lady Verney's Essays.

———

October 7th.

Dear Minnie arrived, also Augusta Freeman and her son Frederick.

A very fine day, with cold wind. Fanny and I drove through Livermere Park and Ampton. Went on with Lady Verney. Read 1 Samuel, chapter xxii.

———

October 8th.

Frederick Freeman went away. Lady Rayleigh (the Dowager) arrived.

Read 1 Samuel, chapters xxiii. xxiv.

Read some of Sainte Beuve.

———

October 9th.

Cecil, the younger (my great nephew) went away, starting for India.

Dear Joanna and Susan Zilari with her, arrived.

———

October 10th.

Wrote a little more of my Reminiscences.

Read 1 Samuel, chapter xxv. Read Sainte Beuve (Causeries du Lundi) on Gibbon.

Looked with Susan through a volume of Lodge's Portraits.

1885.   *October 12th.*

Dear Susan Horner left us—also Susan Zileri.

Went on with my Reminiscences. Read 1 Samuel, chapters xxvi.-xxx.

October 13th.

Read last chapter (xxxi.) of First Book of Samuel. Scott's report on the Mildenhall Rent Audit.

October 15th.

Shut up with a cold, confined to two rooms. Heard from Edward.

Read 2 of Samuel, chapters i. and ii.

October 16th.

A beautiful mild day. I walked with Fanny for a little while in pleasure-ground. I went down to dinner and to the library in evening.

Read 2 Samuel, chapter iii.

October 17th.

Wrote to Edward.

Read some more of Lady Verney, vol. 2

The sad news of the death of Mrs. Chapman.

October 18th.

Read prayers with Fanny, Joanna and Arthur. Fanny read to me a fine sermon of Farrar.

October 19th.

Read *Edinburgh Review* on S. T. Coleridge and on Frederick of Prussia and Louis Fifteenth. Read Matthew, chapter viii. in Greek. Looked over prints in evening with Mrs. Swinton.

October 20th.

I went out for a little while in Pleasure Ground. Mrs. Swinton went away. John Herbert arrived. Read *Edinburgh Review* on Grenville's Memoirs, part 2. Wrote answer to Rev. Mr. Hind of Honington.

October 21st.

Read Monckton Milnes's Monogram on W. S. Landor. Read 2 Samuel, chapter iv. A visit from the darling baby—the little angel.

October 22nd.

Had a pleasant drive in morning with Fanny and Joanna; dear Joanna went away after luncheon— much to our sorrow. Augusta Freeman also went away. Lady Hoste and her daughter-in-law came to tea.

October 23rd.

Lady Charlotte and Lady Octavia Legge and Admiral Spencer arrived, also Clement.

Read Matthew, chapter ix. in Greek. I had a baddish night.

October 24th.

John Herbert went away.

1885.   Read some of Doudan's letters. Scott's report
on the Radical meeting yesterday evening.   Again
a bad night.

_____

Mr. and Mrs. Reginald Talbot arrived.

_____

Finished reading St. Matthew, chapter x. in
Greek. Our guests went to the Bazaar at Bury.
My health much the same.

_____

The John Paleys dined with us.   A rather better
night.

_____

The darling little baby to visit me.   Harry and
Laura dined with us.   A very bad night.

_____

Dismal weather.   Received a letter from Lord
Rosebery, relating to my writings, and asking for a
copy of Memoir of my father—sent him one, and
Fanny wrote to him for me.   Mr. and Mrs.
Reginald Talbot, Charlotte and Octavia Legge,
Admiral Spencer went away.

_____

A fine, bright day.   Went out before the house
with Fanny.   Read chapters ix. and x. of Book 2 of
Samuel.   I felt very weak.

_____

November 4th. 1885.

Received a very courteous and agreeable note from Lord Rosebery about the Memoir of my Father. A slightly better night.

——— —

November 5th.

A long discussion about politics with Fanny and Scott. We had good news of Susan and the others' safe arrival at Florence.

——— —

November 6th.

A visit from the darling baby in high beauty. Laura and her sister Car came to tea with us. I very much *in statu quo.*

——— —

November 7th.

Went out in landau with Fanny and Laura. Lady Muriel Boyle arrived—a pleasant evening with her. A very bad night.

——— —

November 9th.

Mr. Johnston Bevan and his daughter Mabel, also Harry and Laura, dined with us. Began to read Greville's Diary, part 2. A tolerable night.

——— —

November 10th.

Mr. Hind, the clergyman of Honington came hither, and with my consent carried off part of my collection of Mosses to be examined at his home. Lady Muriel Boyle went away. A *particularly* bad night.

——— —

November 12th.

A letter from Edward from London to my great joy. Reading Greville, part 2.

1885.   Weather very gloomy and cold. A really good night—Thank God.

- - - -

A beautiful bright day, quite wintry, but calm and quite beautiful. Had a pleasant drive in the chariot with Fanny. Mr. Hind returned my set of dried Mosses.

- - - -

Sir Andrew Clarke's visit—his code of rules for my health.

- - - -

[Sir Charles was seriously ill in November, and I sent for Sir Andrew Clarke. He said it was a serious case of weakness of the heart, and gave strict regulations as to his diet, and desired that for the present he should keep to one floor, but he was very encouraging and hopeful for the future if Sir Charles followed the rules he prescribed.—F.J.B.]

=====

LETTER.

My dear Edward,

I thank you very heartily for your kind and affectionate letter, which I have received this morning. I had intended to write to you on Christmas eve, but put it off as usual without reason. I am sorry to tell you that Fanny had an unlucky accidental fall in a shop in Bury the other day, by which she hurt her left arm, and gave herself a severe shock; but happily Dr. Macnab 1885. (who has returned from his tour), assures us there is nothing broken, and we may hope she will be quite well in a few days. For myself, I think I must say I am restored to a state of *delicate* health ; that is, I am free from positive malady, but my health is liable to be easily upset; I am certainly much better than when you were last here, and, in particular, I have recovered the power of sleeping very well at night; this is an immense blessing, but I am still kept by the doctors somewhat on the *invalid* list, restricted to *one floor*—not allowed to go up or down stairs, which interferes with my free enjoyment of books, and excludes me absolutely from my *museum*. However, I may well be very thankful that I am so well as I am, and may look forward with hope to another summer.

We have Harry and Laura staying with us at present, and their dear little baby, who is a perfect delight. Arthur too is come back to us from his health-seeking visit to Bath, which seems to have answered well, and Sarah Craig is another visitor. But I cannot write more at present, my head is already tired.

Ever your affectionate brother,

CHARLES J. F. BUNBURY.

1886

JOURNAL.

1886. Confined by medical order (as I have been ever since November 23rd of last year) to certain rooms on first floor in this house of Barton—not being allowed to go up and down stairs.

Mrs. Wilson and her daughters Agnes and Nellie came to luncheon.

Went on reading Memoir of Agassiz, and of Sir Charles Napier.

The Women's* feast.

In statu quo.

Weather mild.

Finished reading the Memoirs of Agassiz, by his widow.

Harry and Laura, with their darling baby, set out for London. I *very* sorry to part with them.

Finished reading Memoir of Sir Charles Napier.

Sarah Craig read to us in evening a part of Macaulay on Lord Chatham —She read very well.

Read some of Plutarch.

* The labourers' wives,

Sent to Mr. Babington my contribution to charities 1886. at Cockfield.

Sarah Craig read to us in evening some more of Macaulay.

LETTER.

Barton, Bury St. Edmund's,
January 8th, 1886.

My dear Edward,

In my letter of the 27th of last month, I told you that Fanny had hurt her arm by a fall in a shop at Bury. That was nearly a fortnight ago, and I confidently hoped (from what the doctors said) that she would be quite well in a few days. But I am sorry to say her arm is (as it seems) worse than when I last wrote. There was nothing broken; but it seems that her arm was severely *sprained*, and that a hurt of that kind is more tedious in healing than a fracture The *left* arm was the one that suffered, so that she is not so much disabled as if it had been the other, but it is very painful at night, and has deprived her of many a night's rest. I am afraid there is rheumatism also complicated with it.

We have been very unlucky as to guests this Christmas; first Frank Lyell and his wife, and now Katharine and Rosamond have been prevented by sudden illness from coming to us: and I am almost afraid of calculating upon any more. It is a trying season. I hope you are bearing it well. I am *in statu quo*—not feeling much the matter with me

1886. at present, but obliged to lead the life of an invalid or a greenhouse plant. But I may well be thankful for being no worse.

Sarah Craig is with us, and no one could be kinder—few pleasanter. But we miss Laura and *the baby*.

Believe me ever,
Your affectionate brother,
CHARLES J. F. BUNBURY.

JOURNAL.

January 9th.

Bitter cold—ground covered with snow. No events.

January 10th.

The sad news of death of dear Cousin Emily.

January 11th.

Wretched weather—a cold thaw—very dark.

Sarah Craig read to us in evening some more of Macaulay on War of Spanish Succession.

January 12th.

Mr. Tanqueray came to us at noon and administered the Communion to Fanny, Sarah Craig and me, in our sitting-room.

LETTER.

My dear Edward,

I am not sure whether you may yet have 1886.
heard of the sad death which has happened among
our nearest relations since I heard from you. Dear
cousin Emily—Emily of Ambleside (I do not know
exactly how else to distinguish her), *is dead*—died
quite suddenly last Friday night. We heard of
it—the mere fact of her death—on Sunday, from
Minnie who had it from Bessy Arran : but we had
no particulars till to-day, and now I do not know
them accurately. It does not appear that she had
any previous signs of illness ; she presided at a
children's party on the Friday evening, and did
not go to bed till half-past eleven, when she said
she was very tired ; in the middle of the night
she rang her bell and complained of violent pain,
and died before the doctor could arrive. I am
much grieved, though for herself it is certainly a
happy end : but I was very fond of her, and she was
always very affectionate to me. I saw her in
London last summer, when she was looking remarkably well, and I thought her a really beautiful
specimen of an old woman ; nobody could have a
better heart.

I am very glad to hear that you have hitherto
borne this winter so well, and heartily hope that
you will continue so to the end.

Ever your affectionate brother,

CHARLES J. F. BUNBURY.

JOURNAL.

1886.   Read some of Plutarch (Cato), and of Woodward's Geology of England and Wales.

Sarah Craig read to us in evening part of Macaulay on Clive.

———

Gave a cheque to Scott for payment of Barton Taxes for year.

Read chapters xiv. and xv. of 2 Samuel.   Reading also Plutarch and Woodward's Geology.

———

Arrival of our dear friends, Minnie, Sarah, Albert and the dear boys, Charlie and Bill — much rejoicing.

———

Captain and Helen Lambart arrived.

I read (not first time) part of Sir Erskine May's Constitutional History of England.

———

Sarah went to London to the wedding of Lady Mabel Gore to Lord Airlie.

———

Sarah Craig went away—she had been a great

comfort to us. Dear Sarah came back from London 1886.
and John Herbert came with her.

I read 2 of Samuel, chapters xvi. and xvii.

——————

January 20th.

The good news of dear Laura's safety, and the
birth of her little boy. Thank God.

A visit from Archdeacon Chapman.

——————

January 21st.

The opening of Parliament.

Again a heavy fall of snow. The *"Queen in
snow !"* built up by Sarah, the boys *et al.*

Read *Quarterly Review* on Don Quixote and on
the House of Condé.

Helen Lambart read to me in evening very
pleasantly.

——————

January 22nd.

Snow continued. Lord John Hervey came to
dinner. Read *The Quarterly* on Burma. Heard
from William Napier from Biarritz.

——————

January 23rd.

The Lambarts went away—very sorry to part
with them. Lady Hoste and Mr. Greene came to
luncheon. Dear Minnie read to me in evening.

——————

January 24th.

Dismal weather—deep snow.

Read prayers with Fanny.

Read *Edinburgh Review* on Victor Hugo.

1886.   January 25th.

A most dismal cold thaw and fog.

John Herbert went away.

Barnardiston and Lady Florence, and one of their
daughters arrived.

A pleasant little dinner.

Dear Minnie read to me in the evening some
of Macaulay (on Sir William Temple).

January 26th.

Business with Scott.

Read *Quarterly* on Church and State.

Dear Minnie again read to me in the evening.

January 27th.

Minnie and Sarah and Albert and the dear
Seymour children went away, to our great sorrow.
The Barnardistons also went away.

Ethel and Ida Wilson, John Herbert, and John
Freeman arrived. Ethel very kindly read to me
in the evening.

January 28th.

Resignation of the Ministry.

Finished reading Book 2 of Samuel.

Went on with Sir Erskine May.

Read Asa Gray on Botany of North America.

January 29th.

News of Lord Stradbroke's death.

Began to read "Mr. Isaacs."

Ethel Wilson read to me in evening very kindly 1886. and very pleasantly.

In statu quo. No events within our own circle.
Heard from Edward a good account of himself.
Reading 1st of Kings; also Broderip's Zool.
Recreat. "Mr. Isaacs."

John Herbert went away.

My 77th birthday. Thanks be to God for his many and great mercies to me.
Dear Mrs. Wilson came to luncheon and tea with us—delightful. Dorothy Hoste and Beatrice Thornhill dined with us.

Reading Forbes's Oriental Memoirs, and Hooker and Thomson's Flora Indica.
Fanny very busy arranging papers.

Reading Forbes, and Hooker, and Thomson.
No events.
Read 1 Kings, chapter ii.
Ethel Wilson read to me in evening, as she has done every day this week—dear girl.

A beautiful bright day, but I still confined to house.

Dear Mrs. Wilson came to luncheon and to tea— very agreeable.

Read over some of my old Journals, and skimmed part of vol. 2, of Forbes's Oriental Memoirs.

————

February 9th.

News of the shameful riots in West of London. Katharine Lyell with Rosamond and little Harry, arrived from London.

————

[Sir Charles was at this time again too ill to write his journal.—F. J. B.]

————

March 11th.

Dear Laura and Harry, with the two darling babies, arrived—all well, and little Cissy more charming than ever.

————

March 12th.

Saw both the dear babies —Cissy lovely— the little boy interesting.

Read chapter xi. of St. John in Greek.   Harry went to Mildenhall for a public meeting.

————

March 13th.

The darlings again.   A visit from Mrs. Horton. Sarah Craig arrived.   Reading Froude's *Oceana*— and a part of Horace Walpole and of Green's History of England.

Staying with us—Harry and Laura and their two dear babies, Ida and Amy Wilson and Sarah Craig.

March 16th.

Deep snow! but it mostly melted before night. Visit from Lady Bristol, Lady Mary and Lord John Hervey—all pleasant.

March 17th.

A visit from Archdeacon Chapman. Mr. and Mrs. Henry Gladstone arrived.

March 18th.

Mr. and Mrs. Henry Gladstone staying with us. Read Rose Kingsley's pretty little book, "The Children of Westminster Abbey."

March 19th.

A bad night—the first for a long time. A change of weather—much milder. Mr. and Mrs. Henry Gladstone went away.

March 20th.

Mild weather, but I still confined to one floor of the house. Dipping into Waterloo and Green's History of England. A pretty good night.

March 22nd.

We heard from Mr. Henry Gladstone, accepting offer of *Vicarage of Barton*. I still confined to one floor of the House.

1886.   March 23rd.

The Bishop of Ely (Lord Alwyne Compton) with Lady Alwyne and Archdeacon Chapman arrived to stay with us for the Confirmation. A large dinner party, at which Harry and Laura appeared for Fanny and me.

———————

                                               March 24th.

Fanny very unwell all day with a bad cold— could not dine with company.

The Bishop and Lady Alwyne and the Archdeacon out all day. They dined with me and Harry and the two Wilson girls—very pleasant.

———————

                                               March 25th.

The Bishop and Lady Alwyne Compton, Archdeacon Chapman, and the Wilson girls all went away. Fanny still very unwell.

———————

                                               March 27th.

Sarah Craig went away. Began to read Harold Finch-Hatton's Australia.

———————

                                               March 29th.

News of resignation of Trevelyan and Chamberlain, and of death of Archbishop Trench.

Harry and Laura with the dear children removed to the Cottage. Nellie and Ida Wilson came to stay with us.

———————

                                               March 30th.

News of death of Sir Henry Taylor.

Signed presentation of Barton Vicarage to Mr. 1886.
Henry Gladstone. Discussion of Barton and Mildenhall rents with Fanny and Scott.

LETTER.

Barton,
March 28th, '86

My dear Edward,

Many thanks for your agreeable letter from Brighton. We have great hopes that Mr. Gladstone will turn out a satisfactory Vicar of Barton. We have lately had the new Bishop of Ely, Lord Alwyne Compton and his wife staying with us for two days, and found them very pleasant; we invited them to stay with us while he was carrying on the Confirmation in this and the neighbouring parishes.

Harry and Laura and their dear children are going to move into the Cottage to-morrow; we shall miss them, and especially I shall miss the darling little girl, my particular pet.

I am very glad you can give a good account of your health; Fanny has had a bad cold, but I hope it is passing off; I am much *in statu quo.*

The prospect of political affairs does indeed appear deplorable, almost hopeless, but I do not at present feel equal to writing anything about them. I wonder what are the books on Irish affairs which you have been reading; I have read none for a long time but Froude and Lecky.

Believe me ever your affectionate brother,

CHARLES J. F. BUNBURY.

JOURNAL.

1886.   Mrs. Wood and Laura came to luncheon with us.
Finished reading Harold Finch-Hatton's Australia.

_____

Read Horace, Sat. L. 1. Sat. 3.

_____

Not well.

_____

My health not satisfactory. Read (not first time)
Froude's "Leaves from African Journal." Played
with the dear children.

_____

News of death of Mr. Forster—a great loss.

_____

William Bruce arrived—a most pleasant evening
with him, Laura, and Harry, and two of the Wilson·
girls.

_____

Much delightful talk with Willie Bruce.—He left
us after luncheon to return to Kensington.   An
uncommonly fine day. Read prayers with Fanny.

_____

The two dear Wilson girls returned to Langham.

* The Book is entitled "Advanced Australia."—F. J. B.

A most beautiful day, but I still confined to the house. Read Prayers with Fanny. Played with the dear children.

April 26th.

A most beautiful day. Reading Lord Beaconsfield's Correspondence. Looked with Fanny through a volume of Curtis's Flora Londinensis.

April 27th.

Mr. Tanqueray came at 3.30, and administered the Sacrament to Fanny and me and Mrs. Wallis. A most beautiful day. Heard from Edward. Laura and Harry went to a ball at Bury.

April 28th.

At last! I was allowed to go down stairs to my study, after being confined to one floor ever since November 23rd, 1885. I enjoyed the change very much. Mr. Henry Gladstone came to stay.

April 29th.

Down stairs to morning room for luncheon, and to my study for afternoon. Mr. Henry Gladstone staying with us.

Began to read Duc d'Aumale on the Princes of Condé.

April 30th,

Dear Mrs. Wilson came to luncheon and tea— delightful. Mr. Henry Gladstone went away.

Began to read Baron Hubner's "Through the British Empire."

1886.   May 1st.

Harry and Laura went up to London for the day
—returning to dinner.

———— —

May 3rd.

The poor little dog Ruby very ill.   I went down
each afternoon to my study.

———— —

May 4th.

A beautiful day.   I had the pleasure—a great one
—of a drive in the open carriage with Fanny and
Laura—First time I out of doors since last November.   Mary Lyell and her children and Edward
arrived.

———— —

May 5th.

A beautiful day.   Enjoyed a drive in the open
carriage with Fanny and Mary Lyell.

———— —

May 6th.

A most beautiful day—perfect summer.
Enjoyed a very pleasant drive with Fanny and
Mary.   Looked into the garden.   Finished reading
vol. 1 of Hubner.

———— —

May 7th.

A very hot day.   Enjoyed a very pleasant drive
with Fanny and Laura.
Looked into the Fern-house.   The dear little dog
quite recovered.

———— —

May 8th.

Another beautiful day.   A pleasant drive and a

pleasant little walk through the arboretum with 1886.
Fanny. The Cottage party dined with us.

May 10th.

Fine, but with cold E. wind. Edward and
Leonard Lyell returned to London. Went through
the garden and arboretum with Fanny and Mary.
The Louis Mallets arrived.

May 11th.

Rain. Read Gladstone's speech (very puzzling)
on the Irish Bill.

May 12th.

Read Lord Hartington's speech (admirable) on
the Irish Union Repeal Bill.
Mr. Livingstone arrived. Rain and very cold
weather.

May 13th.

Very bad weather. Talk with Mr. Livingstone—
he returned to Mildenhall after luncheon. Harry
and Laura dined with us. Signed Mildenhall
Estate account for Mr. Scott.

May 19th.

Rain.
Signed cheques for Fanny.

May 20th.

Our drive baffled by rain.
Signed many cheques for Fanny for the Estate
bills.

1886.   May and Charlotte Egerton arrived. Harry and Laura and Ruth Wood also dined with us.

May 21st.

A most beautiful day.

Had a very pleasant drive in morning with Fanny and May Egerton, and drive in pony-carriage with Fanny in afternoon. Heard nightingales.

May 22nd.

Fine, but a cold rough east wind.

A drive with Fanny and Charlotte Egerton. A visit from Mrs. Bland.

May 24th.

Much rain.

Ethel and Constance Wilson came to stay with us—very pleasant. Read much of Frank Buckland.

May 25th.

May and Charlotte Egerton went away.

Weather showery, but I had a pleasant little drive in the landau with Fanny.

News of Car MacMurdo's engagement.

Finished (second time) Life of Frank Buckland.

May 26th.

Variable weather.

Had a ramble round the garden with Fanny— partly in chair.

The Cottage party dined with us.

The sad news of death of young Norah (Bruce) Whately, three days after birth of her child.

May 28th.

We visited the *Cottage* and saw dear Laura and the two darling children.

Weather rough and blustering.

Read some of the Histoire des Princes de Condé.

May 29th.

Dear Minnie and Sarah arrived—delightful : also John Herbert. We took a rather long drive. The Cottage party dined with us : altogether a delightful party.

May 30th.

Our 42nd wedding day. Thanks to Almighty God for all His goodness to us.

May 31st.

A beautiful day.

Had a delightful drive with Minnie, Sarah and Fanny. Mrs. Wilson came to luncheon—charming.

Went on reading Histoire des Princes de Condé.

June 1st.

Dear Minnie and Sarah and John Herbert went away. A visit from Mrs. John Paley—pleasant.

Fine weather—we had a pleasant drive.

June 2nd.

The *Shrub* — abominable ways — unlucky expedition.—Ego fatigued, but no harm done.

Ethel and Constance Wilson went to Cambridge
and returned late.

————

June 3rd.

Read prayers with Fanny. Fanny had caught
cold, and neither of us could go out.

I went on with Princes de Condé, and read part
of Alfred de Vigny's Cinq-Mars.

————

June 4th.

Beautiful bright weather with cold wind.

Fanny could not go out—I walked a little with
Laura.

Readings as before.

Mr. Henry Gladstone arrived.

————

June 5th.

Jack Freeman arrived, fresh from Egypt.

Fanny recovered from her cold—we walked in the
garden.

Weather continuing bright and cold, with east or
north wind.

————

June 7th.

A most beautiful day.

We had a very enjoyable drive through Livermere
and Ampton.

Jack Freeman went away after breakfast, and Mr.
Gladstone after luncheon.

Read Wells on Physical Geography of Brazil— 1886.
Geographical Society's Journal.

June 8th.

The *great division* in the House of Commons—
majority of 30 against Gladstone (341-311).

Our dear Mrs. Storrs arrived—as charming as
ever.

June 9th.

Weather as before — bright sun and cold east
wind. Had a very pleasant drive with Fanny and
Mrs. Storrs. Dear Kate Hoare arrived while we
were at dinner.

June 10th.

Dear Mrs. Storrs went away—Ethel Wilson also.

I received a botanical letter from Mr. Hind of
Honington: began an answer to him.

June 11th.

A beautiful day. A delightful drive with Fanny
and Kate Hoare.

Finished letter to Mr. Hind on botany.

My account of Charles' last illness, which I was advised to write soon after his death.—F. J. B.

1886.   June 11th, 1886.—We had Kate Hoare staying with us from the evening of Wednesday the 9th, which Charles particularly enjoyed. On this day, Friday, June 11th, Charles, Kate, and I took a drive. We drove to Rougham to see the rhododendrons, which Charles admired very much. When we got to the crossing of the roads—one going up to the Mill, the other to Thurston, I asked him which way he would like to go ; and he decided to go by Thurston to Pakenham.

We called at Nether Hall, but Lady Hoste was out. He particularly enjoyed this drive ; it was a beautiful day, and the air very soft and balmy, and he delighted in seeing the fresh, spring vegetation. When we came home, he rested a little while in his study, and then we three had a cheerful luncheon together. After luncheon he rested in his study till four o'clock, when we three took a walk in the pleasure grounds, resting sometimes on the chair near the house. We went to tea at five o'clock, in the library ; at half-past five or quarter-to-six, he was carried to his bedroom, where he sat reading till time for dressing for dinner.

When I went to see him, he gave me a letter he wished sent to the post, written on Botany, to a Mr. Hind, clergyman at Honington. I rather reproached him for this, but he said he had taken two

days to write it. We had a very happy dinner, and 1886.
when he had rested for half-an-hour after it, Kate
and I went up to him, and I read Gilbert Gurney to
him, which rather amused him. He went to bed
about 10.30. p.m., and he said "that he had
enjoyed his day so much."

On Saturday, June 12th, he did not take quite so
much for breakfast as usual, that is to say, he left
two of his rusks ; but when I asked him an hour
or two afterwards how he felt, he said, "quite well."

We had been generally reading in bed together,
or rather I read to him, a chapter of The Acts ; the
Collect for the week, the prayer beginning "Almighty
"God to whom all hearts be opened ;" also the one
beginning "Almighty God, who art more ready to
"hear than we to pray ;" also the prayer blessing
"our dear absent ones," and very often praying that
he might be restored to health and strength, that
we might lead a life of usefulness; the Thanksgiving
and the Lord's Prayer.

We were also reading the "Pilgrim's Progress,"*
we had read part 1, and two chapters of part 2. I
think we read all that morning.

Kate Hoare left us that morning after breakfast,
she bid Charles good-bye very affectionately, saying
that she "rejoiced to see him looking so well, and
"hoped that they would find him as well if she and
"her husband paid us a visit in the autumn." As I
was talking to Mr. Scott, he saw Charles going

---

* He remarked to me that his mother had never appreciated the works of the
Puritans, and had never given him this book to read, but that he was very
much struck with the force of its language.

1886. down stairs. I soon followed, and at twelve we started
for our morning's drive, and were out nearly an
hour. Laura, her brother, and Mr. Scott were at
the door, and Laura thought him looking very well.
We called at Mr. Scott's to enquire what had been
done for poor Mrs. Dorling.* Mrs. Scott spoke to
us at the carriage, and told Mr. Scott afterwards
that she thought Sir Charles looking pale and
yellow. He did not enjoy the drive so much as he
did the previous day, and I said, in fun, that it was
because Kate Hoare was not with him : but the
weather was not so genial. After we came home,
we took a walk in the pleasure-grounds, and he told
me the names of some of the trees on the lawn ; it
was threatening to rain, and I asked him whether he
would like to go home, or whether he would rather
go on further, he said he would like to go to the
garden to see his ferns, if it was prudent, not on
account of the rain, but rather on account of his
strength ; I told him he was the best judge of that,
so we went on, and when we got to the fern-house,
he sat down for two minutes, and then he walked
round the fern house while I was sitting down, being
tired myself. He especially remarked the "Wood-
wardia" and the New Zealand "Tree Ferns." We
then went back to his study to rest before luncheon,
he sitting in his arm chair, and I lying beside
him on the sofa asleep. At two o'clock we went
to luncheon, and he appeared quite well, he took
some soup, but afterwards refused a little trout
which had been dressed up from what we had had

* She had lost her husband.

the day before, I urged him to take it, he complied 1886. as he generally did with anything I asked him to do, but said it was dry and poured some white sauce over it. My cook told me afterwards that the fish was quite fresh. He then took a good quantity of asparagus; he refused a little cup pudding, the first time I ever saw him do so; but when I expostulated, he remarked, "You cannot expect me "at my age, to eat with the appetite of a young man." I called for my maid Jeanie Davidson, and gave her the pudding and some strawberries to take to some poor people: we then rose from luncheon, and Charles, as usual, went to ring the bell for the servants, when he became very giddy, and I saw him clinging to the table beside the fireplace. I ran and caught him before he fell, and Jeanie, who was on the stairs, joined me, and we placed him on a chair, as he could not stand; the men servants soon came and carried him in the carrying chair to the sofa in his study; this was a little before three o'clock: he became violently sick and retched a great deal. I sent for the nurse, who was in the house, and telegraphed for the doctor. Dr. Macnab came about half-past three. He had been given brandy and water, but the retching was so excessive that it had exhausted him much, and he was deadly pale. Dr. Macnab left him about four o'clock, and he got to sleep about half-past four and slept for an hour, when he again began to retch: however, Dr. Macnab's medicine had then arrived, which soon relieved him; he lay on the sofa till half-past nine, when the men carried him to his bedroom; he

1886. was soon placed in his bed, from which he never
rose again. The nurse said it was not paralysis, as
he could move his legs perfectly in bed.

June 13th, Whitsunday.

Dr. Lucas, who stayed all night, gave a satisfactory account of him. The next morning they fed
him entirely on chicken broth with milk and
whiskey. I didn't read to him that day, as quiet-
ness was important for him. In the afternoon,
Dr. Macnab came, who cheered me by saying that
Charles had turned the corner, and the wave had
passed. In the evening, his cousin Jack Freeman,
(a doctor) arrived, and he also cheered me very
much by giving his opinion of him.

I telegraphed for Agnes Fincham to come from
Mildenhall. In the evening she was sitting by the
bedroom fire, but afterwards went to the bed to him,
when he said to her " Oh, I am so glad to see you,
" I thought you were a stranger."

Monday, June 14th.

Jack Freeman left us in the morning, but before
going, gave me a very favourable account of Charles,
but he said he would require two or three weeks to
regain the strength he had lost from this attack;
this depressed me a good deal, so I told Dr. Lucas
what he had said, when he came in the morning.
Dr. Lucas gave also a favourable account of Charles
—and added, " I hope we shall see him out in a
" very few days, sitting in that chair," pointing to
one that had lately been placed in the pleasure
grounds.

Dr. Lucas said that I might read to him, and that the next day he might read to himself. I read to him that morning most of the 2nd chapter of the Acts (the description of the day of Pentecost), up to the 40th verse, and at the end he said, "What a "very fine chapter !" Then we read our usual prayers. Then I continued reading to him, " Gilbert Gurney." While I was reading, he twice asked, " What sound was that ? " I never could ascertain whether this was a sound in his own ears, or whether it was the sound of the drums and fifes of the Whit Monday party ;— probably the latter, as it never occurred again. In the afternoon Dr. Macnab saw him, and was much disappointed from the favourable account given by Jack Freeman and Dr. Lucas. Charles' speech was much affected, and this distressed him very much, but Dr. Macnab said he would soon get over that ; he told me he must not be read to. Dr. Lucas said it was quite unnecessary for him to stay all night, as he was so much better.

Tuesday, June 15th.

In the night he had a fainting fit, which frightened the nurse, and we sent for the doctor. Dr. Lucas came, and slept all night with him, and did not think so favourably of him.

In the course of the morning he asked me to let him have a book to read, I told him that Dr. Macnab did not wish him to read ; but to keep quiet : Charles told me I knew better than Dr. Macnab, and when Dr. Macnab arrived in the after-

1886. noon, he again asked him to allow him to read, so Dr. Macnab half consented and took up a book, which was Mrs. Oliphant's novel "The Curate in Charge," with a good print, and offered it to him, Charles said, "Dull," and asked for a Waverley Novel ; Dr. Macnab said he might have one. When I came back to him after the doctor's visit, he said, "When may I have the book?" I said "now," and asked him which he would like, and he said, "Waverley," so I went down to his study and brought him up the 1st vol. After tea I went up to him again, and he asked me to read to him, which I consented to do, he showed me the place, then took the book out of my hand and showed me the exact place ; before reading he asked me if it would tire me. I read two paragraphs to him, then I read him a letter from Kate Hoare, which interested him very much with the account of the approaching weddings of her two brothers, but he became a little confused, and thought it was "Truey" who was going to be married ; when I stood at the bottom of the bed he called to me and said " Is it Truey that " is going to be married ? "

Wednesday, 16th June.

Early in the morning Dr. Lucas saw him, and he was quite rational and cheerful, and asked for Ruby who jumped up on his bed, and Dr. Lucas brought her to him to pat.

I think he had a fainting fit this morning, and some time afterwards he talked incessantly—so unlike himself—sometimes he did not know me, but

latterly he was quite himself, and when I asked 1886.
if he would like to see his brother Edward, he said,
"he would like it," and when Edward came, he
kissed his hand over and over again, and Edward
kissed him, both his hands and his lips. I think he
must have known he was going then, but I did
not think so at the time. He then saw Katharine
and recognised her, and kissed her. During Wednesday afternoon, I was alone with him, and hearing
Bower outside, I asked him to come and help
me to raise him, he recognised him directly, and
asked him how he was.

Mrs. Wilson called and sent her love to him;
he said, "I am much obliged; give my love to her."
During this afternoon, I read a good part of the 5th
chapter of Matthew, in a monotonous voice to
soothe him; when I had done, he said, "it was
long."

Towards the evening, he became much worse, and
all, including Dr. Lucas, assembled round him,
expecting it to be his last. I knelt by his side, and
kept my hands over his, which were at first very
chilly, but they became very warm—his pulse became
gradually weaker and weaker, at last two beats in a
minute. I was always asking if he was gone, when
an extraordinary rally took place, and Dr. Lucas
said, "his pulse is up to 50." This was at 2 o'clock
a.m. on Thursday morning, the 17th June; all then
left the room, except me, nurse Sutcliffe, and Agnes
Fincham. I sat with my hands over his till half-
past six in the morning, seeing him sleep sweetly:
he then woke with a most heavenly, joyous smile,

1886. and he kissed my hand over and over again, and offered his mouth to kiss me, and asked for Ruby, who generally came to us in the morning. I sent for her, but he was asleep before she arrived. Dr. Macnab called that morning, and was astonished at his rally; he was quite himself in every way. It was this morning he said he was afraid I had not money enough, and I said I had plenty. In the afternoon he was more excited, and while Katharine was sitting by his side, he seemed to be making a long address like a public speech, but she could not make out many words—*opportunity* was one of the words she heard.

Later on, after he had spoken some time, Katharine wishing to interrupt him, called Harry, who was with me in the next room, and he said, "Are you well, Harry?" twice; and Harry said, "Quite well, Uncle Charles." Then he said, "How "is Laura?" Harry said, "Quite well, would you "like to see her?" "Yes, I should," he said: " Then she said, "How do you do, Uncle Charles?" " Cissy is quite well." And he said "I am so glad."

It was after this, I think, that he wanted to get up and write. Fincham begged him not. Then he said, "Lady Bunbury will write for me." I came to him and said, "You know, darling, I can write all you want," and he was quite satisfied, but never told me what to write.

Late in the evening, before he settled for the night, he took some nourishment. I asked him if he liked it, he said, "Yes;" and told me he was "very comfortable."

He had always me or Fincham or Jeanie with 1886.
him, besides the nurse. His nurse gave him about
5 o'clock a.m. some whiskey and milk, which un-
luckily I did not see him take, as I was asleep
beside him. She gave it to him lying down, and he
swallowed it well, and then fell into a sleep from
which he never woke ; no speaking either from
me or Dr. Macnab, could arouse him, and about
5 in the evening, he breathed his last without a
struggle—so peacefully.

EXTRACTS FROM LETTERS,

WRITTEN BY FRIENDS AFTER HIS DEATH.

"Oh! what a beautiful delicate nature his was! I never knew any one just like him, so richly endowed and yet so modest, and exquisitely humble! So pure in heart—so courteous to all, whoever they might be!—ready to learn from every inferior; and if all these graces and gifts were to be seen by those who only saw him casually, how beautiful must the *inner* life have been, known only to God, with whom one felt he must have such a habit of deep and reverent communion. 'Blessed are the pure in heart, for they shall see God.' These words repeat themselves to me when I think of DEAR Sir Charles. How my husband *loved him, revered him, prized* every hour with him."

———

"Quiet and retiring as he was, I think he exercised an amount of quiet influence for good, the extent of which some will very likely never have fully realized until now, that it is to a certain extent withdrawn; and yet is not, and please God, can never be, wholly lost; for the life of such a high principled Christian gentleman leaves a memory behind, by which even 'being dead he yet speaketh.'"

———

"With great sorrow, with sympathy most sincere

for you, I learn that all is over at Barton, and that I shall never see again in life, though I shall always have him in remembrance,— one of the kindest, most cultivated and intelligent, most courteous and most just in word and deed of all the men I have ever known."

———————

"Your dear husband was so loved and valued by us; there was no one whose society was so charming to us, and his great qualities of *heart* as well as of head, are precious to look back upon."

———————

"Certainly his life was a very noble one, so unselfish, so gentle, so unassuming, with his great gifts held with such attractive humility. The perfect gentleman, the cultivated student, and the modest Christian seem to combine in him. His death has cast a great gloom over this parish, where he has been so much revered and beloved."

———————

"Having been admitted, through your kind hospitality, to what I shall always consider the great privilege of having known Sir Charles, I can truly say that the charm of his conversation drawing so quietly upon his most abundant store of knowledge, and his beautiful courtesy, will always live in my recollection."

———————

" The world has lost one of the purest and noblest

characters that were ever born into it. Dear
Charles! his was a very rare nature, but one whose
influence will not easily pass away from the memory
of his friends."

———

"I have lost in your brother Charles a much-loved
friend, whom I regret most deeply, and of whose
mental and moral qualities I had the highest
opinion."

———

"To have been honoured by his affection and
friendship is a thing to be thankful for all one's life.
There surely was never a truer friend than he was.
His beautiful humility, which so often shamed our
younger and more arrogant judgments, his unselfishness—his chivalrous courtesy—his great wisdom and
temperateness on all questions that came under
discussion—his wide learning and his never-failing
goodness, combined to make what has always
seemed to me a unique character. We shall never
see his like again. But I think we must be a little
the better for having known such a man."

———

"He was one who combined great intellectual
powers with a remarkable sense of duty. It is not
often that one sees the character of a scholar and a
country gentleman so thoroughly united."

———

"We both were struck by his remarkable gentleness and courtesy, and by the great modesty that

accompanied his very unusual talents. It was impossible to be in your house without feeling how
blessed a union yours was, and what an influence for
good was shed around you."

––––––

" Though the world seems so much the poorer for
the loss of those who are so good and noble and
pure being taken, it cannot *really* be so,—their
memory lives ; they are gone to higher, nobler work
elsewhere. I do not think any life is unfinished
in God's eyes or in His scheme of goodness and
wisdom. Is it not a comforting thought, dear
Fanny, that dearest Charles was one of those whose
love and trust and faith were as pure as when
he was a child ? whose deep knowledge and wisdom
never made him otherwise than humble at the con-
templation of the highest wisdom. I always think
of him as one of the 'Saints of God.' "

––––––

" I feel it is not for me to attempt an estimate of
his manifold worth, but you may not be displeased
to know how deeply even a comparative stranger
was impressed by the genial benignity which seemed
to radiate from your brother's very presence, and
the beauteous modesty of his intellect, learning !—
last and rarest grace."

––––––

" Surely no man had more friends, and all his
friends loved and admired him for himself and his
own great qualities. I do not know any one who

used so modestly his unrivalled stores of knowledge, who intruded them so little, but who was so instantly ready to apply them in assisting to remove any doubt or difficulty which others felt. Never appealed to in vain, his accuracy of memory was something wonderful, and yet his corrections of others was so gently done that it scarcely sounded like a difference."

———

"Dear Sir Charles! I cannot say how I valued and shall ever value his affection and friendship, of which this is a fresh and most touching proof.* As you kindly wish me to express a preference, I should prefer either a volume of poetry or biography, but I shall be more than satisfied with whatever you select. I mention poetry as it is specially connected with him in my mind from his so often quoting passages appropriate to any subject under discussion, and it was always a great pleasure to see his face light up as the lines came into his mind. But, as I said before, please choose whatever you would prefer yourself."

———

"Dear Lady Bunbury, it is so sad to me to think of you alone, and of Barton without that dear, gentle, refined master, who made it so peculiarly reposeful and lovely; he was so good to me, so kind, and gave me always such an unaffectedly genial welcome, that I grieve to think I shall never more see him here, and he has left no one like him —so reserved, and yet so affectionately kind."

* From a lady, to whom he left a book.

" May I say it is a personal grief to me to know that I shall see him no more ; but I shall always feel it a privilege to have known him ; and the beauty of his gentleness and humility, combined with his great knowledge, made a character one is proud to have known, and to claim some kinship with."

"You know how we all loved and revered dear Uncle Charles. His gentle, beautiful life will have done good to all who have had the privilege of knowing him."

" I shall never forget his kindly courtesy, his tender unselfishness, and the modesty of a true scholarship which made him what he was to us who knew and prized him as a neighbour and friend. We are all the poorer by such a friend, the richer by a memory."

" It is impossible for us without profound sadness, to think of the loss of one for whom we entertained the sincerest regard, and whose very distinguished and attractive qualities, we have so long appreciated and admired—the central figure of a circle in which we have passed so many happy days."

"I wish particularly to thank you for your kindness in enabling me to see dear Sir Charles almost up to the very close of his peaceful and beautiful life. I have lived many years at your gate, and I am most thankful to believe that during all those years, our relationship to each other as landlord to tenant, and neighbour to neighbour, have, without a single exception, been completely happy ; and I feel sure that one whose object has always been to dispense ' peace and happiness, truth and justice,' has gone where those inestimable blessings are for ever enjoyed in their most complete perfection."

———

" Dear cousin Charles ! it is very beautiful to me to recal his gentleness and wisdom—his delight in all good things, and in the happiness of those around him. The influence of a good man's life—of those ' little nameless, unremembered acts of kindness and of love,' works on quietly in the hearts of all that knew him for many a year, thank God ! and cannot be destroyed by death."

———

" I feel that I have one *real friend* the less on earth since he was taken away from us. I valued him greatly for his great qualities both of mind and heart, and always loved his truly christian simplicity and piety. I had a very real regard and affection for him."

———

From the Obituary Notices of the Proceedings of the Royal Society. Vol. 46.

" Sir Charles James Fox Bunbury, Bart.* was born at Messina, in Sicily, February 4th, 1809, where his father, General Sir Henry Bunbury, was at that time Quartermaster-General. His mother was a daughter of General Fox, then commanding in the Mediterranean, and a niece of the celebrated states-

---

* The author of this notice is Sir Joseph Dalton Hooker, K.C.S.I., C.B.

man. To these gifted parents Sir Charles owed
his early love and his knowledge of arts, literature
and science, and especially of natural history, ac-
complishments which he cultivated throughout life
with disinterested zeal; and thanks to his extraordinary memory, his accuracy was as remarkable
as were the extent and variety of his information.

"After completing his education at Trinity College,
Cambridge, Mr. Bunbury visited Brazil and the
River Plate, whither he was attracted by the fact of
his uncle, Mr. Fox, himself an ardent collector of
plants being Minister at Monte Video. This was
followed by a voyage to South Africa, where another
uncle, General Sir George Napier, was Governor of
the Cape Colony; and in 1853 he accompanied Sir
Charles Lyell to Madeira and Teneriffe. In all
these countries Sir Charles Bunbury made extended
excursions, observing diligently, and collecting
assiduously, though travelling as an amateur rather
than a scientific naturalist. The results of these
journeys are full of interest to the botanist, zoologist
and geologist; they are published in various
scientific periodicals, and in a ' Visit to the Cape,'
which appeared in 1847. Especially valuable are
the botanical observations made in South Africa
and South America, which deal with the broad
features of a vegetation known previously only in
detail. They are brought together in a volume
published shortly before his death, entitled
' Botanical Fragments.'

"It is however by his researches in vegetable
palæontology that Sir Charles Bunbury is best

known as a scientific man. To this subject his attention was more immediately drawn, through his connection by marriage with Sir Charles Lyell, and his most valuable contributions to it may be said to be ancillary to Sir Charles's investigations into the coal measures of British North America and the United States; they appeared in the form of a succession of communications to the Geological Society of London, between 1846 and 1861, and are printed in that Society's Journal. He also wrote on the Carboniferous flora of the Tarentaise, on the Anthracites of Savoy, on the Jurassic flora of Yorkshire, on the Fossil plants of Nagpur in the Deccan Peninsula, and of the Island of Madeira. Under this head, too, should be recorded the great services he rendered to palæontology, by classifying and naming the Carboniferous fossils in the Museum of the Geological Society (of which Society he was Foreign Secretary from 1847 to 1853). This collection was for many years the only one of its kind in England available to geologists or botanists.

" In 1844, Mr. Bunbury married Frances Joanna, daughter of Leonard Horner, Esq., F.R.S., and sister to Lady Lyell, who survived him. In 1860 he succeeded, through the death of his father, to the baronetage, and removed from the Manor House, Mildenhall, to the family seat, Barton Hall, Bury St. Edmund's, where he died, June 18th, 1886, leaving no descendants. He was a Fellow of the Linnean and Geological Societies, as well as of the Royal Society, into which he was elected in 1851."

JOSEPH D. HOOKER.

Milton Keynes UK
Ingram Content Group UK Ltd.
UKHW041522181024
449640UK00009B/139